# Tubular wire welding

# Tubular wire welding

DAVID WIDGERY

WOODHEAD PUBLISHING LIMITED

Cambridge England

Published by Abington Publishing
Abington Hall, Abington, Cambridge CB1 6AH, England

First published 1994, Woodhead Publishing Ltd

British Library Cataloguing in Publication Data
A catalogue record for this book is available from the British Library.

ISBN 1 85573 088 X

Designed by Geoff Green (text) and Chris Feely (jacket).
Typeset by Best-set Typesetter Ltd, Hong Kong.

# Contents

*Preface*     vii

**1 Why flux-cored wire?**     1
     Productivity     3
     Ease of use     5
     Quality     6
     The process     8

**2 Types of wire**     12
     Basic wires     13
     Rutile wires     16
     Metal-cored wires     18
     Self-shielded wires     25
     Alloyed wires     29
     Sheath materials     31
     Production methods     31
     Wire section types     35
     Wire lubrication     37

**3 Planning for productivity**     39
     Joint preparation     39
     Wire and gas selection     43
     Burn-off and deposition rates     45
     Welding speed     47
     Pulsed arc welding     48
     Mechanisation and robotics     50

4   Planning for quality                                    52
        Lack of fusion                                       52
        Porosity                                             55
        Hot cracking                                         59
        Cold cracking                                        60
        Metallurgical quality of the weld                    65
        Microstructure and mechanical properties             68
        Post-weld heat treatment                             72
        Troubleshooting                                      72
        Monitoring the process                               75
        Training and technique                               76

5   Process economics                                       85
        Costing a joint                                      92
        Measures of productivity                             96

6   Equipment requirements                                 100
        Power sources                                       100
        Wire feeders                                        105
        Welding guns                                        108

7   Standards                                              111
        Gas standards                                       122

8   Applications and consumable selection                  124

9   Health and safety                                      134
        Radiation                                           135
        Fumes                                               136
        Gases                                               138

    References                                              140

    Index                                                   144

# Preface

The use of tubular, or, as it is often known, flux-cored wire has grown dramatically in the last thirty years and in more recent times has continued to expand despite a declining total market for steel and welding consumables. However, few textbooks have covered the subject in any depth and the picture emerging from technical seminars has been fragmented. Unlike longer established products for manual metal-arc and submerged arc welding, many tubular wires are made with formulations and process technology that are not in the public domain or are buried in patents with which users are not familiar. At the same time, some fairly extravagant claims have been made for the process which are difficult to evaluate without a working knowledge of the principles behind it.

This book aims to give fabricators an unvarnished account of what tubular wires can do and how they do it. It has grown out of lectures given to MSc students of welding and the questions they asked. Not all the detail of what goes into wires will be essential to every user: it is included as a defence, should it be needed, against suppliers who try to blind purchasers with science. This is a versatile and productive process which needs no hyperbole to encourage its use.

The decision to use a particular welding process involves many parts of an organisation: welders on the shop floor, welding engineers, designers and general management can all contribute to it. It is hoped that they will all find useful information here. Where some prior knowledge of welding technology is assumed, references to more general textbooks are given.

In 'Self-Shielded Arc Welding', my former colleague Dr Tad Boniszewski made a plea for self-shielded welding to be treated as a process in its own right, rather than as a sub-division of flux-cored arc welding. While agreeing that self-shielded welding deserves the separate metallurgical treatment that he has elegantly provided, it is argued here that the processes have enough in common to warrant dealing with both types of wire when it comes to the principles of planning for productivity, controlling quality and evaluating economic benefits.

Authors of academic papers are expected to provide references or experimental data to support any technical claim they make. This is not so easy for a writer drawing on industrial research not published elsewhere. Even though it is generally lack of time rather than more commercial considerations that prevents publication of the details of such research, these would not be appropriate in a book intended for a general readership. Readers will therefore be met with some *ex cathedra* assertions of a more or less sweeping character. Where possible, fuller references are given when controversial views are discussed or where the reader can be directed towards useful background information.

My thanks are due to the Esab Group for permission to go into print. They have observed the production of the book from a proper distance and cannot be held responsible for the opinions expressed in it or for any errors that may have crept in, but have been most helpful in providing illustrations and the opportunity to meet many people involved in making and using flux-cored wire. Thanks are particularly due to colleagues who may recognise their own ideas uncredited in the text. Thanks are finally due to my family for their forbearance during the writing of the book.

# CHAPTER 1

# Why flux-cored wire?

Metal-arc welding, in which an arc is struck between a consumable metal electrode and a workpiece, appeared in the final decade of the last century and covered electrodes, bearing a coating of flux and alloying elements, in the first decade of this century. In Oscar Kjellberg's remarkably comprehensive patent of 1912, he says 'The rod may be of any desired shape, or a tube may be used, wherein the substances that are to give the weld the proper chemical composition are contained, either as a dried paste or well mixed together'.[1]

It has been said that the life cycle of welding products is long compared with that of other products of twentieth century technology. After eighty years which have seen the rise and fall of the dirigible, the valve radio and the vinyl gramophone record, welding with tubular wires is still regarded in some quarters as innovative and possibly dangerous. The following chapters will attempt to dispel this notion.

Kjellberg's patent identified a valuable feature of covered electrodes. If the coating melts at a slightly higher temperature than the core, a ceramic cup is formed on the end of the rod which both directs the arc, making vertical and overhead welding possible, and partially protects the metal within from the atmosphere. If the fluxing agents are contained inside a tube, both these advantages are lost: hence Kjellberg proposed putting alloying elements in the core of the tube but preferred to leave the flux outside. In later years ingenious formulists succeeded in getting wires to turn themselves inside out as they melted, but this required a mastery of melting behaviour and surface tension which was not immediately available.

As long as electrodes were supplied in cut lengths for manual use with an electrode holder, there was little economic advantage to be gained from the tubular form. Nevertheless, many thousands of tonnes of tubular electrodes were supplied in the 1920s and 1930s. They were used in the manufacture of ships, bridges, machinery, railway vehicles and a range of products where toughness requirements were modest by present standards. But the real benefits would come when it became possible to spool and feed continuous lengths of tubular wire.

Semi-automatic gas-shielded welding with solid wire was described in 1926[2] in a form close to that which is familiar today, albeit with some different shielding gases. As improved manufacturing techniques allowed the production of tubular electrodes in smaller diameters, it became practicable to feed them through similar semi-automatic equipment and so flux-cored arc welding in its modern form evolved in the years leading up to the Second World War.

In recent years, pressure to increase productivity and reduce costs has been the main driving force behind the adoption of flux-cored wire by fabricators. Other things being equal, the ability to weld continuously without stopping to change electrodes must speed up the process and it is a matter of surprise that semi-automatic welding in general came into use so slowly. It was not until the 1960s, nearly forty years after it was first demonstrated, that the method began to make any serious challenge to the established manual metal-arc (MMA) process and it was another twenty years before, at least in Europe, it drew level in terms of annual usage. The introduction of flux-cored wire in Europe was partly a response to two waves of competitive pressure: from the USA in the 1960s and 1970s, and from Japan in the 1980s. By understanding what lies behind the timorous response of industry to a process which appears to offer dramatic increases in productivity without sacrifice of quality, it may be possible to exorcise some ghosts of the past and suggest ways of improving matters in the future.

Some of the resistance to semi-automatic welding in general arises from the increased capital cost of equipment. It will be seen in Chapter 5 that this consideration should not weigh heavily with serious users of welding in the Western world. Semi-automatic welding with solid wires gained an early reputation for producing lack-of-fusion defects, or cold laps, which were often difficult to detect by radiography, the only non-destructive examination technique available at the time. Flux-cored wire welding has been unfairly tarred with the same brush, though much less susceptible to the problem. For added confidence, modern ultrasonic inspection can detect cold laps if they should occur.

Some reservations about flux-cored wire do no doubt stem from the form of the product itself. If part of the coating is missing from a stick electrode, that is immediately visible to the user before an arc is struck. The filling of a tubular wire is hidden from view and open to suspicion: certainly, concern about how the presence of a uniform fill can be guaranteed is a recurrent theme in the auditing of suppliers of the product by users and the question will be discussed in later chapters. Manufacturers, it must be said, have not always been as open in discussing the formulation of tubular wires as they have with stick electrodes. A recent European standardisation committee writing a standard for flux-cored wires sought analogies for the chemical descriptions such as 'basic-rutile' or 'cellulose-rutile' found in the MMA standard, but concepts such as 'barium fluoride-lithium ferrate' were not thought familiar enough to users to be included.

These legitimate concerns of potential flux-cored wire users are generally dispelled by familiarity with the process and confidence grows with use. The three cardinal virtues of the process then become apparent: productivity, ease of use and quality.

### Productivity

The welding processes with which flux-cored wire must mainly compete are MMA and semi-automatic welding with solid wire (MIG/MAG). The advantages of a continuous process are clear. Changing electrodes not only takes time, it tends to impose its own slow rhythm on the proceedings, providing natural breaks which extend beyond the time strictly necessary for the change. Duty cycles have often been much lower than management suspected or at least admitted and costings based on optimistic figures have done fabricators no service. A 1971 survey[3] used annual electrode consumption per welder to arrive at an overall

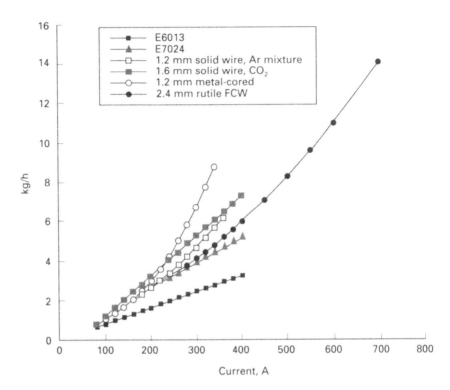

**1.1** Deposition rates of MMA electrodes (E6013 and E7024) and tubular wires.

duty cycle of 12%, while contemporary agreements in the ship-building industry limited use of basic MMA electrodes to 5 rods per welder per hour. Conversion to semi-automatic welding does not by itself guarantee an improvement in duty cycle, but taken in conjunction with other necessary changes in working practice and industrial relations, it can open up that possibility.

Even at the same duty cycle as MMA welding, flux-cored wire can give major productivity improvements because it can operate at much higher current densities. Productivity will be dealt with in more detail in Chapter 3, but Fig.1.1 shows that in terms of deposition rates there is rarely any contest between MMA and flux-cored arc welding.

When comparing MIG/MAG welding with solid wire and with flux-cored wire, deposition data alone is less convincing. Solid wires when run at very high current densities, for example 350–600 A/mm², can certainly achieve deposition rates comparable with those of flux-cored wires and from time to time interest in this technique is revived by the introduction of a new gas mixture said to be especially suited to it. Given a wire quality that will allow the very high wire feed speeds needed, up to 46 m/min being quoted, and given a mechanised or robotic process so that the operator is not exposed to the high levels of radiation and ozone generated, high deposition-rate MIG appears to be viable. However, the use of a larger diameter flux-cored wire running, if necessary, at a slightly higher current but with a cooler-running gas, often results in greater process tolerance and comfort. Where extreme deposition rates are not needed, flux-cored wire is frequently chosen in preference to a solid wire for its improved penetration and bead profile, which can make reliable operation possible at higher travel speeds.

Productivity depends on more than just good deposition rates, and some of the ways of planning for productivity are dealt with in Chapter 3. Of special importance is the balance between the energy used in melting the consumable and that available to penetrate and fuse the parent material. Many processes, including some using flux-cored wire, achieve high deposition rates at the expense of poor penetration. This can make it difficult to control the pool in vertical welding or to produce defect-free welds in the flat or horizontal-vertical positions. Fortunately, the technology of flux-cored wire offers the formulist several means of varying the relative melting rates of wire and base material, provided the user does not insist on kilograms per hour per amp as the sole criterion of productivity.

Productivity is also a function of all the things that go on when the arc is not burning: deslagging, grinding of stop-start areas, changing electrodes or wire reels and so on. These too must be brought under control and the use of properly chosen tubular wire can maximise time and so reduce costs.

### Ease of use

The argument in favour of welding consumables that are easy to use has often met with suspicion, never more so than when solid wire MIG welding was introduced. Within a short time of striking his first arc, a MIG welder can produce good-looking welds. Closer inspection, however, can sometimes reveal the weld metal merely resting in the preparation, on which the original machining marks may still be visible.

The fact is that all welding processes need proper training for the operators, not only in the manipulation of the consumable and the arc, but in some of the theory which enables them to set up the equipment correctly and to detect when things are going wrong. In this respect, solid wire MIG welding can be quite demanding because the amount of heat available to melt the parent metal when welding with dip transfer is not great and avoiding cold laps takes constant vigilance. Most tubular wires offer some improved resistance to cold laps compared with solid wires, but for positional welding the rutile types have two advantages. Metal transfer takes place by spray transfer at all currents, so that the arc plays continuously on the parent material, and the availability of stiffer slags allows higher currents to be used without loss of control of the weld pool.

Of all the arguments for using flux-cored wire, ease of use is the least convincing in print and perhaps the most convincing on the shop floor. It may simply be noted that when large numbers of welders have been needed quickly for critical work, as for example when yards were being set up to build the Forties platforms in the North Sea, training unskilled personnel from scratch in the use of flux-cored wire has often proved the most practical proposition.

### Quality

If MMA welding had not preceded all the other arc welding processes, perceptions of the relative quality of semi-automatic welding could have been different. The introduction of a process where stop/start positions have often to be ground out and electrodes have to be baked before use and kept in heated quivers might have been met with the sort of enthusiasm that would have greeted Sir Walter Raleigh if he had returned with tobacco to the world of the 1990s. Nevertheless, the long and successful history of stick electrodes and the persistence with which their inherent problems have been overcome have led to their present predominance in many quality-sensitive applications such as pressure vessel construction where flux-cored wires are still not fully accepted.

In fact, there should be no reason to exclude tubular wires from any application where stick electrodes have proved acceptable. Modern manufacturing and monitoring techniques are quite capable of controlling the filling ratio of the wire and eliminating the old bugbear of encountering a stretch of empty tube. Early wires were produced without much attempt to control or remove drawing soaps, so porosity came to be regarded as a fact of life with many of them; today, tubular wires probably have a better record in this respect than stick electrodes.

Perhaps the most serious area of concern about the quality of welds made with flux-cored wire is that of achieving sufficient penetration to avoid lack-of-fusion defects. The absence of a flux cup to direct the arc when using flux-cored wire has already been mentioned. At the time when semi-automatic welding with both solid and flux-cored wires was being introduced, the most popular stick electrodes were the cellulosic type which used additional propulsive forces to achieve fearsome penetration. By contrast, wires with no external coating or quantities of hydrogen available must rely on electromagnetic forces to shape the arc and project droplets into the pool.

Developers of flux-cored wires have risen to the challenge and found ways of improving their penetration. Arc modifiers in the powder promote the formation of the conical arc which allows electromagnetic forces to accelerate the droplets away from the wire tip. In dip transfer, this cannot happen because the droplets are too bulky and the distance for acceleration too short, but careful control of the heat balance and of the wetting properties of the metal and slag ensure that, even here, penetration is significantly better than for solid wire. With the additional benefits of modern power sources, weld defects related to lack of penetration should cause no more concern with tubular wire than with any other welding process.

According to International Standard ISO 2560, MMA electrodes may be defined as 'low hydrogen' if they deposit weld metal containing no more than $15\,ml/100\,g$ of hydrogen. An increasing proportion of electrodes meet this criterion, but there is no reason why all flux-cored wires should not do so. As standards encourage manufacturers to make and verify more ambitious claims for their products, it will be seen that many tubular wires are capable of giving hydrogen levels below 5 ml/100 g of deposited metal. This must further reassure welding engineers considering changing to the process.

## The process

A brief description of the process of welding with tubular wires here may serve as a reminder of the many ways in which it resembles welding with solid wire as well as pointing to some important differences. Numerous publications describe the general features of semi-automatic welding more comprehensively than space here permits. Variations in terminology have survived continuing attempts by national and international bodies at standardisation and these will be mentioned.

Semi-automatic welding involves the use of a continuous electrode which is fed through a hand-held torch or gun. The former term was not sanctioned in British Standard 499 Part 1 of 1991 but was simultaneously approved by the International Electrotechnical Commission and so appeared in the 1992 supplement to BS 499. At the end of the torch the wire passes through a contact tip (BS 499: contact tube) through which the electrical power is transmitted. Where gas shielding is required, a cylindrical nozzle often known as the gas shroud surrounds the contact tip. The welding wire is generally pushed from a wire feed unit, which may or may not be mounted on the power source, through a flexible conduit of spirally wound steel to the gun. This is bound together with the power cable and gas supply tube in a single outer sheath. Figure 1.2 shows some typical equipment for welding with tubular wire.

Power is provided by a transformer-rectifier or other power source, typically at between 15 and 35 V. In gas-shielded welding with solid wire, at low currents and voltages the wire regularly makes contact with the molten weld pool and during the resulting short circuit the tip of the wire melts. As the molten zone grows, surface tension pulls the droplet into the pool. When the narrowing bridge of liquid between the droplet and the wire tip finally ruptures, an arc is re-established. The wire again moves forward through the arc, being heated as it advances, until it once more touches the pool, the arc is extinguished and the process begins again. This happens up to 100 times a second and the process is

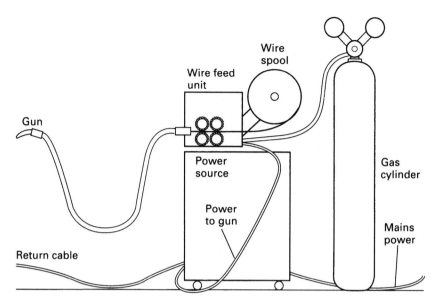

**1.2** Equipment for welding with tubular wire.

known as dip transfer or short-circuiting transfer, Fig. 1.3. When solid wires are used to weld vertically, dip transfer has to be used to limit the heat input and the fluidity of the weld pool.

At higher voltages and currents, enough energy is available to maintain a longer arc and to melt off droplets from the wire tip before short circuiting occurs. As current and voltage are increased, the droplets are at first larger than the wire diameter and the transfer is coarse and globular, but then the droplets become finer. Eventually, transfer takes place through a steady stream of fine droplets pouring off a pointed wire tip and pro-pelled into the pool by a plasma jet. This is known as spray transfer.

A similar transition in behaviour is found with some types of tubular wire. At this point it may be noted that until recently, the expression 'flux-cored wire' was in universal use on both sides of the Atlantic, though it failed to gain entry to the British standard on welding terminology, BS 499. When the first British standard on these products was proposed in the 1980s, it was pointed out, perhaps pedantically, that a few wires existed which contained no fluxing agents at all, having an entirely metallic filling. British

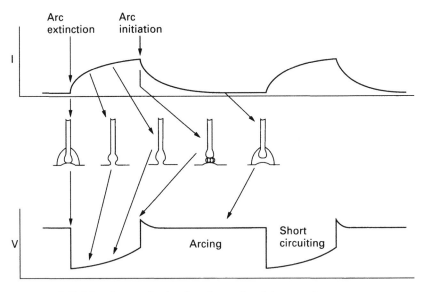

**1.3** Metal transfer in the short-circuiting mode.

Standards therefore describe the products as cored or tubular cored electrodes, the term 'flux-cored wire' being 'deprecated' (BS 499: Part 1, 1991). The European Standardisation Committee (CEN) has so far followed suit. It remains to be seen whether common usage can be stamped out so easily, especially when the American Welding Society (AWS) shows no signs of abandoning it, and the expression will appear from time to time hereafter for the sake of better mutual understanding.

Tubular wires with basic or metallic fillings operate in dip transfer at low currents like solid wires and metal-cored wires move into spray transfer at high currents, but basic wires cannot be made to show a true spray transfer. Wires with a rutile core, on the other hand, work in something very like spray transfer at all currents. These differences will be discussed further in the next chapter.

Unlike commercial solid wires, some types of tubular wire are designed not to need an external gas shield. This greatly simplifies the equipment requirements, described in Chapter 6, but there are penalties to be paid in metallurgical properties and productivity. When shielding gases are used, the most common ones are carbon dioxide, $CO_2$, and mixtures of this with argon, usually abbreviated Ar in this context, and sometimes helium, He.

Tubular wires are available for open arc (gas-shielded or self-shielded, as opposed to submerged arc) welding in a much wider range of sizes than solid wires, at least from 0.6 to 3.2 mm diameter. Solid wires of such large sizes would give rather globular and unpleasant metal transfer unless very high currents were used to force them into the spray mode. Tubular wires are also used for submerged arc welding in which the arc is struck beneath a layer of granular flux, and in that case may be up to 4 mm in diameter.

The distance from the end of the contact tube to the end of the wire in semi-automatic welding is important in determining the type of metal transfer and the deposition rate but different terms are used in describing it. Common usage speaks of 'wire stickout' but BS 499 favours 'electrode extension'. The term 'stand-off' is sometimes used to mean the distance from the contact tip to the workpiece surface, which will often approximate to the electrode extension since when the process is correctly set up, the arc penetrates the plate to leave the wire tip roughly in the plane of the original plate surface.

# CHAPTER 2

# Types of wire

Tubular welding wires consist of a metallic sheath in the form of a simple or complex tube containing a powder filling which is usually partly metallic and partly non-metallic. The material of the sheath does not need to reproduce exactly the required weld metal composition, since the alloying elements can conveniently be added to the core of the wire. Where the total content of alloying elements is to be high, however, restrictions of space in the tube may dictate the use of an alloyed sheath.

In gas-shielded flux-cored wire, the fluxing agents or slag formers which constitute the non-metallic part of the powder fill have to perform several functions. It has been proposed, and is implied by the use of such names as 'Dual-Shield', that the flux provides a secondary shielding action in addition to that of the shielding gas. In reality, this rôle has been exaggerated for ferritic materials, as the later development of metal-cored wires has shown. What the flux can do is to control or adjust the oxygen content of the weld metal, raising it or lowering it according to the needs of the application.

In common with steelmaking slags, some welding slags are able to remove certain impurities such as sulphur from the molten metal, but with good quality modern steels this ability is less needed than it once was. More important are the physical characteristics of the slag, which can mould or support the weld metal or help it to wet the base material.

Some non-metallic components of the powder are not there as slag formers but to stabilise the arc or to control the burn-off characteristics of the wire. Such materials may be present even in the so-called 'metal-cored' wires. In flux-cored wires, arc stabilisers must be selected so that any residues remaining do not adversely affect the slag.

Wires designed for use without an external gas shield must contain other constituents to protect the weld metal from atmospheric contamination. These include volatile minerals and metals, and materials which break down to produce both, as well as deoxidants and nitride formers intended to capture any oxygen and nitrogen which manage to penetrate the vapour shield.

### Basic wires

Early flux-cored wires were produced by developers whose experience was in stick electrodes, so it was natural that they should use slag systems which had been proved in MMA welding. The earliest MMA slags were of the basic type and these could be transferred with little modification to tubular wires.

MMA electrodes have to provide their own shielding so limestone or other forms of calcium carbonate were added to the coating to liberate carbon dioxide in the arc. To stick the powder on to the core rod, silicate solutions were used and the residue of these helped to flux the residual calcium oxide, lowering its melting point. As an additional fluxing agent, calcium fluoride in the form of fluorspar was used. By this means the slag melting point was adjusted to just below that of the weld metal so that it would not be trapped as the weld cooled. Calcium fluoride also volatilised in the arc, providing an effective screen against atmospheric gases.

Because lime-fluorspar slag systems were developed in the first place for their ability to shield the transferring metal, it may seem surprising that they should have become so widely used in a gas-shielded process. In fact, early versions of the basic wires were often sold for use with or without gas at the user's option, and this is reflected in the original version of classification EXXT-5 of the American Welding Society (AWS).[4] Today, the practice of using such wires without gas shielding survives significantly only in Eastern Europe, but other virtues ensure their continued popularity for gas-shielded welding. Foremost of these is their ability to produce excellent mechanical properties in the deposit.

One of the main sources of oxygen in weld metal is the decomposition of silica in the slag. In the presence of an excess of calcium ions derived from calcium oxide and calcium fluoride, silica forms silicate ions which have a much lower tendency to dissociate. Many publications[5] have demonstrated the lowering of weld oxygen contents as the basicity ratio $(CaO+CaF_2)/SiO_2$ in the slag increases to a value of about 2. Oxygen is relatively insoluble in steel, so as the weld metal cools towards solidification, it comes out of solution in the form of small, usually spherical oxide inclusions. Oxygen contents are higher in weld metals than they are in bulk steel because the rapid solidification does not allow inclusions time to float to the surface as they can in ladles or ingots. Inclusions have an adverse effect on weld toughness, the more so when alloying is used to increase the tensile strength of the steel.

Sulphur, also relatively insoluble in steel, is the other main source of non-metallic inclusions in welds. Indeed, a simple calculation suggests that to a first approximation, the total volume fraction of inclusions vf is given by the equation vf $(\%) = 5.5(wt\%O+wt\%S)$. In addition to promoting low oxygen levels, basic slags also desulphurise the molten metal as they do in steelmaking. The very clean weld deposits thus achieved give the consumable designer the maximum flexibility to achieve optimum combinations of strength and toughness.

When steels had higher impurity levels than they do today, weld metal solidification cracking, sometimes loosely called hot cracking, was feared by welding engineers who found that the ability of basic slags to cope with sulphur often gave them a weapon against this defect. Phosphorus, the other element mainly responsible for solidification cracking, cannot be removed by welding slags since these do not provide a sufficiently oxidising environment. Today, solidification cracking in structural steels is rarely a problem though the use of basic consumables may be advisable when repairing older structures or when joining engineering steels which were not designed with weldability as a prime consideration.

In MMA welding, the achievement of low deposit hydrogen levels effectively requires the use of basic electrodes. Other types of electrode often use moisture in the coating to improve their running characteristics and to provide a degree of shielding against atmospheric nitrogen, but both limestone and fluorspar in different ways not only shield the arc from the atmosphere but also inhibit hydrogen pickup. The same applies to flux-cored wires. These are not formulated to have moisture deliberately present, so it is normal for all types of flux-cored wire to meet the 15 ml/100 g criterion for a 'controlled hydrogen' electrode, but the presence of fluorspar in particular in the core gives the weld a degree of tolerance to adventitious hydrogen in the arc, whether this comes from residues from the production process or from contamination of the workpiece by oil or other organic matter. Most basic flux-cored wires should be able to give deposited metal hydrogen levels below 5 ml/100 g.

An incidental benefit of using basic wires is their ability to weld over pre-fabrication primers better than other types. This arises mainly because of the fluidity of basic slags, which allow gases evolved during welding to pass through the slag rather than to form bubbles at the metal-slag interface. The relatively high level of deoxidation of basic wires also helps when the primer is of the iron oxide type and the ability of the slag to remove sulphur is helpful on primers which use sulphur compounds as 'extenders'.

For positional welding, basic wires are used in short-circuiting or dip transfer, and perform best in this mode with electrode negative, which reduces the size of the droplets and minimises the formation of a 'boot' on the end of the wire. Welding sets designed originally for solid wire MIG welding sometimes have wire feed units which are not insulated from the power source, making a change to negative polarity impracticable. This was in the past especially prevalent in certain parts of Europe. For these markets, wires have been developed which will run with electrode positive, but the compromises involved have so far resulted in poorer performance in positional welding.

Basic wires thus offer a combination of excellent mechanical properties, low hydrogen contents and tolerance to plate condition even including primed surfaces which have allowed them to continue almost unchanged in concept and formulation for 25 years or more. On the debit side, their fluid slags and reluctance to achieve spray transfer has meant that positional welding has traditionally had to be undertaken with a short-circuiting arc, to the accompaniment of some spatter and a risk of lack-of-fusion defects or cold laps. Chapter 6 will show how the newest welding equipment, especially if it has pulsing facilities, can go some way towards overcoming these disadvantages. In the meantime, rutile wires have been developed to take over in many areas from their basic counterparts.

### Rutile wires

Rutile, a form of titanium dioxide, became a popular base for stick electrode coatings in the 1930s. It allowed the melting point and viscosity of the slag to be controlled over a much wider range than was available with basic slags, so it was possible to make electrodes with stiff slags for vertical welding or fluid slags for high speed welding in the flat position. Furthermore, titanium is a good arc stabiliser and is often added to basic flux systems, in either metallic or mineral form, to give a smoother arc. These benefits are seen equally when rutile is used in flux-cored wires.

Titanium dioxide is a very stable compound that would contribute little oxygen to the weld, so when it is combined with basic components the resulting product retains many of the characteristics of a fully basic system. Unfortunately, these can include excessive fluidity and a tendency to globular metal transfer, so true rutile-basic hybrids have to be approached with caution. For the best weldability, the presence of siliceous or acid material is helpful.

To achieve stable metal transfer as a fine spray, the surface energy of the droplets must be kept low. The easiest way to do this is to allow some oxidation of the droplet surface, and we have seen that basic components inhibit transfer of oxygen from the slag while acid components, principally silica, promote this. Users of rutile flux-cored wires have come to expect that perfect spray transfer will be achieved over a wide range of operating conditions, so developers are reluctant to use slag systems that will reduce weld oxygen contents much below 650 ppm. To reach levels of toughness suitable for offshore applications with this constraint was a challenge which took some years to meet.

Rutile melts at between 1700 and 1800 °C, so some fluxing agents are needed to bring its freezing point down to the 1200 °C or so required for a welding slag. This in itself is not a problem, since many minerals will form low melting point eutectics with rutile. The difficulty for the electrode designer lies in the short time available for the powders to melt.

If the distance from the contact tip to the melted end of the wire, the stickout length, is, say, 20 mm and the wire is fed at 6 m/min or 100 mm/s, a point on the wire will spend only 0.2 s in traversing the stickout. Heat is being generated by resistive heating during this time in the sheath only, and must be conducted into the core, ideally in time for the core and sheath to melt almost simultaneously. High speed photography shows that when this happens, the slag can wrap itself round the metal droplets. When the core has not melted in time, a cone of powder can be seen hanging from the wire tip like ash on a cigarette, preventing axial transfer of the droplets. Solid, powdered minerals are not good conductors of heat, and if the core consists entirely of materials with high melting points the fact that, once melted, the mixture will freeze at a relatively low temperature may not be enough to allow good welding properties. The solution is to include among the powders at least one with a melting point below that of the sheath, which can help to carry the heat through the core and to dissolve the rutile and other minerals of high melting point.

As early as the mid-1950s, cored wire manufacturers were using synthetic titanates of sodium, potassium and manganese to ensure good melting. Some wires even dispensed with rutile in mineral form, relying solely on synthetic titanates to form the slag. The melting characteristics of such wires are still remembered with affection, unlike the porosity from which they tended to suffer because many titanates readily absorb moisture. High cost was another disadvantage, and when wire sizes started to decrease in the 1970s, so that heat had less far to travel in the core, many manufacturers were glad to reduce or abandon their usage of pre-fused additives.

At about this time, the presence of sodium and potassium titanates in rutile wires was noticed by astute marketing experts, who coined the expressions 'rutile-basic' and 'hybrid' for the new generation of E71T-1 all-positional wires which were just starting to appear. Most of the new wires actually contained less synthetic titanates and were less basic than their downhand-only E70T-1 predecessors – and were all the better for it in terms of moisture pickup and weld hydrogen content. Nevertheless, promoters of the new products succeeded in attaching to the E71T-1 wires something of the perceived quality of basic MMA electrodes. The toughness of the new wires was good partly because the residual impurities in the steel strip were getting lower all the time and partly because deoxidation practice was better understood and boron was usually added to the wires.

Just as there was no real quantum jump in basicity between the old E70T-1 wires and the new E71T-1 types, so the positional welding capability improved more because of the availability of smaller wire diameters, especially 1.2 mm, than because of any breakthrough in slag formulation. Some of the old formulae perform as well as the all-positional wires if made in small diameters and formulists would generally have difficulty in sorting a random assortment of formulae into E70T-1 and E71T-1 types. This is not to devalue the remarkable progress made in flux-cored wire performance over the last 30 years, but it does suggest that the progress has resulted from many detailed improvements in both the design and manufacture of wires, rather than from any single dramatic leap forward. Figure 2.1 shows the excellent weld profile that can be achieved with a current 1.2 mm rutile wire running under 20% Ar, 80% $CO_2$ shielding.

The user of rutile wires should thus not be distracted by claims as to their semi-basic or 'hybrid' qualities, but should choose a wire strictly on performance. Some have particularly attractive welding characteristics, whether by virtue of their formulation or of their method of manufacture, while others give more reliable low-temperature toughness. Some wires deposit weld metal with a very low hydrogen content, but it will be seen in Chapter 4 how important it is to compare like with like when dealing with hydrogen measurements and how difficult that has been until now using published data. New standards, described in Chapter 7, may help in the future.

### Metal-cored wires

Tubular wires whose core consisted only of metallic ingredients, with no minerals, were used for hardfacing early in the history of the process, but for joining wires it was felt that a slag was needed to improve weldability and protect the weld pool. A metal-cored wire for joining was described in a patent filed in 1957,[6] but its only *raison d'être* seemed to be that solid wires of the same composition were difficult to obtain. Seventeen years later, a more detailed patent application from the same company[7] allowed for the inclusion of small amounts of non-metallic constituents but

**2.1** Vertical weld made with all-positional rutile wire showing flat profile.

was still modest in the claims made, which centred around better deposition rates and weld profiles compared with conventional flux-cored wire. Given the improvements in flux-cored wire since then, these claims would hardly stand up today. More significantly, the patent mentions the avoidance of finger penetration when using argon-rich shielding gas mixtures and, in passing, the possibility of designing wires to weld under $CO_2$ with a more stable arc than solid wires. These indeed are for modern users two of the most compelling reasons for using metal-cored wire.

The first period of major growth in the use of metal-cored wire came in the 1970s, when $CO_2$ was still the predominant shielding gas in solid wire MIG welding. A large part of the European welding industry was owned by gas suppliers and the expansion of oxygen steelmaking had led to the production, as a by-product of oxygen distillation, of significant tonnages of argon for the first time. Meanwhile, though this is difficult to believe today, the supply of solid welding wire was failing to meet the demand. Commercial imperatives thus fuelled the development of metal-cored wire with argon-rich shielding gases as a high productivity, low spatter process. Users of solid wire with $CO_2$ could not fail to be impressed, though solid wire users who had already converted to argon mixtures might have taken more convincing of the wilder claims made by promoters of metal-cored wires.

After an initial burst of enthusiasm, metal-cored wires did suffer from a degree of over-selling, based on misconceptions which manufacturers were sometimes reluctant to dispel. The proposition was that in tubular wires, current is carried only by the sheath, not by the core. The reduced area of metal carrying the current results in more $I^2R$ heating in the stickout, and the higher wire burn-off rate which this causes leads to higher productivity.

In fact, tubular wires are made with a sheath of rimming or at any rate very soft steel, while solid wires contain more than 1% manganese and almost as much silicon and so work-harden at a faster rate, increasing their electrical resistivity. It is not a foregone conclusion therefore that tubular wires will have a higher resistance. Japanese work[8] has shown that, at least for temperatures in the ferrite range, tubular wires of moderate wall thickness can have less electrical resistance per unit length than solid wires of the same diameter and this is one reason why metal-cored wires often run at lower voltages than solid wires. When a user substitutes a metal-cored wire for a solid one and has to increase the wire feed speed to maintain the same welding current, this may be not because the deposition rate has gone up but simply because the wire density is lower so a greater length of wire is needed per gram of deposited metal.

Deposition rates can be increased by reducing the tube wall thickness (increasing fill percentage) and this was often done in the early days of metal-cored wires. As mentioned earlier though, this changes the balance between the energy used to melt the wire and that available to melt the parent material, which leads to reduced penetration. Welding engineers are now abandoning the crude measure of kilograms per hour per amp as the sole criterion of productivity, allowing tubular wire makers to reduce filling ratios and burn-off rates and to provide more attractive welding characteristics. Higher currents may be needed to produce the same fillet sizes, but welders who cannot see the ammeter some-times believe that the current has been reduced, such is the improvement in comfort and controllability.

If arguments based on burn-off rates of metal-cored wires are now seen as mostly irrelevant, there remain the sounder if more modest claims of the original patent. 'Finger penetration' in solid wire welding with argon-based gases puts restrictions on joint fitup and seam tracking accuracy that can be relaxed when metal-cored wires, with their better underbead profile, are substituted, Fig. 2.2. This is particularly important in mechanised and robotic welding where there is no welder to compensate for small tracking errors. Travel speeds can often be raised 30% or more by the use of metal-cored wire, a significant benefit in a capital-intensive application.

When the time came to replace MMA welding with a semi-automatic process, the shipyards of Japan and Finland in par-ticular favoured metal-cored over solid wire, at least as much for reasons of quality as for the productivity benefits which certainly followed. Shipyards are now the largest users of metal-cored wires. A particular advantage in this case is the wires' relative tolerance to welding over pre-fabrication primers, for which some wires have indeed been specially developed. Rutile wires are at a dis-advantage here because of their rather thick and stiff slags which prevent the gaseous decomposition products of primers from escaping.

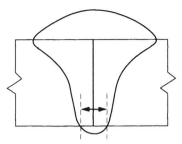 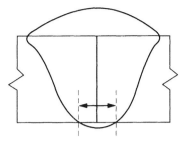

Solid wire: narrow penetration
profile gives poor tolerance
to tracking errors

Metal-cored wire: wider penetration
and better tolerance

**2.2** Penetration profiles of solid and metal-cored wires.

In Japan, welding companies did not fall under the control of gas companies as they did in Europe, and $CO_2$ welding continued through the 1970s and 1980s almost unchallenged by processes based on argon. It was natural, therefore, that the first metal-cored wires to exploit commercially the claim of improved performance on $CO_2$ came not from the original patentee, BOC, but from Japan. For every tonne of wire consumed, a welder uses in round terms half a tonne of gas, so although the cost of gas may not be great in relative terms, it is still an item in the cost equation and $CO_2$ costs, in the UK, about 60% less than 80% Ar, 20% $CO_2$. More importantly, there are some technical advantages to be gained with $CO_2$. Less radiant heat is generated, torches run cooler, welders are more comfortable and less ozone is formed. Because $CO_2$ is denser than argon, the gas shield is more resistant to disruption by draughts. Weld hydrogen levels can be up to 30% lower with $CO_2$ than with 80% Ar, 20% $CO_2$. For all these reasons, metal-cored wires designed for use with $CO_2$ have appeared throughout Europe and the USA in recent years. The technology behind them is interesting, involving complex cocktails of arc stabilisers which typically constitute only a small fraction of one per cent of the wire weight, together with manufacturing processes which must ensure that each droplet forming on the wire tip will contain all the components of the cocktail in the correct proportions. The development of metal-cored wires for $CO_2$ may prove to be one of the more significant commercial developments of the late 1980s.

The 1974 patent on metal-cored wires covered fill percentages from 22 to 45 with non-metallic materials comprising 0.25 to 4% of the core. It made clear that as well as acting as arc stabilisers, some of the mineral constituents could behave as acid or basic slag components, exerting a metallurgical influence just as they do at much higher levels in conventional flux-cored wires. A family of wires was developed in Europe using basic compounds such as calcium fluoride to reduce oxygen and hydrogen contents in the weld metal. Eventually, alternative ways of achieving the same result were found for most wire types, but fluorides persisted in those for welding higher strength steels. It came as a surprise, therefore, to see the same concept propounded as a novelty in an American patent of 1989.[9]

These wires still produce welds with minimal slag cover, Fig. 2.3, but at the top end of the range of mineral contents quoted, wires may be produced which give a full, if thin, slag cover on the deposit. Most metal-cored wires may be used, if not quite as successfully as basic wires, to weld over pre-fabrication primers, and this ability is preserved with slag levels up to about 4% of the total wire weight if the slag is fluid enough not to impede the passage of gaseous by-products of primer decomposition.

On shot-blasted or ground plate, metal-cored wires made with minimal slag-formers produce so little slag that it is normally possible to make a three-run fillet without deslagging between runs. Even more runs are possible in some cases. Thus what appears to be a very simple class of products still presents the user with a range of choices; some wires are formulated with minimum slag as the main design criterion, others for the best performance on primer, still others for high toughness or optimum weldability on $CO_2$. The latter options will be identified in the new draft European standard on tubular wires, prEN 758, which is likely to come into force in 1994.

Most metal-cored wires have a preferred polarity, often electrode negative for baked wires and positive for unbaked types. A degree of surface oxidation as found on baked wires helps to stabilise the cathode (negative pole) and prevent the arc root from climbing up the wire in search of a stable site. Where both polarities are recommended, electrode negative is usually chosen for positional work but electrode positive may give an advantage in welding on contaminated or painted plate.

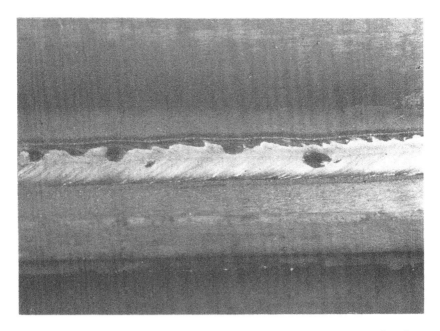

**2.3** Surface of weld bead made with metal-cored wire, showing small slag islands.

Even when basic components are not present in metal-cored wires, many are capable of matching basic wires in terms of weld metal hydrogen levels. The oxide film which surrounds the droplet, which is provided by designers mainly to encourage spray transfer, also inhibits hydrogen pickup. For this reason, some of the earliest high strength wires were of the metal-cored type. Other alloy types available include nickel-bearing wires for low temperature toughness, Cr-Mo wires for creep-resisting applications and copper bearing wires for weathering resistance, although the latter are being increasingly replaced by 1% Ni or 2.5% Ni types.

Metal-cored tubular wires even in their standard form may be used for submerged arc welding given a suitable flux and equipment and with some modification they can be made to perform even better. Some productivity advantages are described in the next chapter, but the ease with which a small quantity may be produced, rather than the minimum 20 tonne or so needed to have a special cast from which to make solid wire, is likely to make tubular submerged arc wires increasingly important.

## Self-shielded wires

To provide shielding gas at the welding torch can be difficult, cumbersome or expensive. In many parts of the world, gases are not readily available. When welding on tall structures, heavy gas bottles are a potential hazard. In quite light breezes, shielding gas is blown away. For all these reasons, self-shielded tubular wires have been developed in parallel with gas-shielded types.[10]

The first demonstrations of self-shielded wires as we know them today came almost simultaneously in the USA and Russia in the late 1950s.[11,12] Over the next thirty years, the product has been developed to compete with stick electrodes and with gas-shielded wires in many areas, but especially in structural work up to and including offshore fabrication.

Attempts to weld ferritic steels without adequate protection from the atmosphere lead to porous deposits because nitrogen dissolves in the liquid metal much more than in the solid, and cannot escape in time as the pool freezes. Oxygen also contributes to porosity. In designing wires to run without external shielding, the consumable developer has two options: he can provide materials which vaporise or dissociate to form a physical barrier against the atmosphere, or he can use nitride formers and deoxidants which will render harmless any nitrogen and oxygen which do find their way into the pool. In practice, both approaches are used together.

Calcium fluoride or fluorspar has been a popular base for self-shielded wires. Having a boiling point of about 2500 °C, a temperature plausibly claimed to be reached or exceeded at the droplet surface, it generates a voluminous vapour blanket around the wire tip. There is also a second benefit: fluorides, for reasons which are still being debated, seem to inhibit the absorption of gases at a steel surface by more than a simple reduction in partial pressure would predict. There are, however, some disadvantages in the use of fluorides. Fluorspar in particular has an adverse effect on arc stability and metal transfer. The high rate of vaporisation generates fume, and condensates form on exposed surfaces near the weld. Nevertheless, calcium fluoride remains the main non-metallic constituent of many self-shielded wires.

It might seem logical to use carbonates in self-shielded wires. They are widely employed in covered electrodes, and the $CO_2$ which is liberated when carbonates break down is a popular shielding gas in semi-automatic welding. Unfortunately, it turns out that the production of too much $CO_2$ within the tube increases spatter and, in extreme cases, can force open the seam of the tube as it passes into the arc. Thus the proportion of carbonates in the slag-formers tends to be lower in flux-cored wires than in MMA electrode coatings.

Lime and fluorspar are used together in basic stick electrodes and in basic flux-cored wires. AWS Specification A5.20 of 1969 described the use of basic E70T-5 wires with or without external shielding gas. The vapour produced by the combination of lime and fluorspar, when used with a long enough stickout to heat the powder fill to a high temperature before it reaches the wire tip, can protect the weld metal without the need for large amounts of extra deoxidants or nitride formers. However, the weldability of the T-5 products when used with long stickouts is not attractive by modern standards and their use without gas is not mentioned in the latest edition of A5-20.

Self-shielded wires have instead developed with a combination of vapour shielding from non-metallic materials, metal vapour shielding and powerful nitride formers.[13] Several metals can perform both of the latter functions. Aluminium melts at 660 °C, boils at 2467 °C and reacts to form stable oxides and nitrides. Virtually all the self-shielded wires suitable for multi-pass welding use it as their main nitride former. Metallurgists are aware that as a strong ferrite former also, it can inhibit the development of austenite to such an extent that weld metals with excess aluminium may show no microstructural transformation, and hence no grain refinement, as they cool. To prevent this, carbon is often added as an economical and effective austenite former. Users accustomed to weld metals with typically less than 0.1% carbon should not be alarmed when they encounter a self-shielded type with up to 0.3%; the effects of the carbon and the aluminium are carefully counterbalanced so that excessive hardenability and hot cracking susceptibility are avoided.

Magnesium is another volatile element which is also a nitride former, while titanium and zirconium are strong nitride formers but do not contribute to the vapour shield. Lithium, on the other hand, helps to provide a voluminous vapour blanket. Some patents have claimed that weakly bound compounds of lithium, such as the carbonate or lithium ferrate, must be reduced by strong deoxidants to yield lithium for this purpose,[13] but this now appears too restrictive a view of the ways in which lithium may be exploited.

Perhaps the most controversial as well as among the most useful components of self-shielded wires are compounds of barium. Barium carbonate was included in E70T-5 wires intended for gasless use because it had a higher decomposition temperature than calcium carbonate and was less prone to explosive breakdown in the stickout, but it soon became clear that barium compounds had other helpful effects. The main benefit was their ability to sustain very short arcs, a result of the low work function of barium compounds. Typically, if a calcium fluoride-aluminium wire needs 22 arc volts at a given current to prevent 'stubbing-in' of the wire and extinctions of the arc, the same wire made with barium fluoride might need only 13 or 14 volts. Two advantages follow from this. When welding in position, lowering the arc energy and hence the wire burn-off rate for a given current gives the welder added control over the pool. Self-shielded wires developed specially for pipe welding, which deposit only 1.4 kg/h of metal at a current of 240 A, succeed largely because of their barium content. The other great benefit of a low arc voltage, and hence a short arc, in self-shielded welding is the reduced opportunity for nitrogen pickup by the droplet. Most of the wires designed for high toughness, and all of those used in offshore construction for the North Sea, are of the barium-containing type. Some barium compounds are toxic, which has led to concern over the use of others in welding: this will be discussed further in Chapter 9.

Rutile is such a widely used ingredient of welding slags that it may seem surprising not to find it more commonly in self-shielded wires. Two difficulties conspire to exclude it. First, the strong deoxidants necessary for these products reduce metallic titanium from the slag. This then produces fine carbide and nitride precipitates in the weld metal which both strengthen and embrittle it. Only if the wire is designed for welding thin material with modest ductility and toughness requirements is this acceptable, and some commercial wires are indeed made on this basis. Secondly, reducing $TiO_2$ yields, as an intermediate product, $TiO$, which has a lattice parameter close to that of the ferrite matrix of the underlying weld metal and causes the slag to stick to the metal by a process very like that used in vitreous enamelling. Even welding over a root run made with a rutile stick electrode can cause slag removal problems with self-shielded wires. Those wires designed using rutile contain a relatively large amount of the mineral and are only lightly deoxidised, which gives acceptable slag removal.

The core of a self-shielded wire has extra work to do compared with that of a gas-shielded type, and given that the production of voluminous slag and gaseous protection for the metal is one of its functions, many of the constituents used are of low bulk density. This may lead to difficulties in packing them all into the limited space available, so every ingredient must justify its place in the mixture, if possible by combining two jobs in one. So fluorspar forms both a slag and a vapour shield, while aluminium is a nitride-former and deoxidant as well as generating vapours in the form of both the metal and a volatile sub-oxide. Nowhere is the designers' ingenuity in this area better seen than in the use of the pre-fused or sintered components of self-shielded wires. The need for such materials was pointed out in relation to rutile gas-shielded wires, where their main purpose is to improve melting by transferring heat from the sheath to the core as the wire passes through the stickout. This rôle is even more essential in gasless wires, whose efficient shielding depends on the fluxing materials being already in a molten state at the wire tip and able to wrap the droplet in a coating of slag. Another use of pre-fused material, as in rutile wires, is to enable alkali metals to be added in a non-hygroscopic form. In this case lithium ferrate is the example most often quoted, the idea being that this is reduced to metallic lithium and iron by the strong deoxidants present. Pre-fusing core materials can also increase their density, so making it easier to fit them into the limited space available, and improve their particle size distribution, so helping the smooth flow of powder into the tube during manufacture.

The remarkable dedication of a small number of developers over thirty years has resulted in the ability of self-shielded wires to tackle far more welding applications than could have been foreseen at the outset, from offshore construction to pipe laying and motor vehicle manufacture. Where no gas supply exists, or where welding must be undertaken in adverse weather conditions, they remain unchallenged. On the other hand, a price has to be paid in deposition efficiency for filling the wire with extra shielding material and in productivity for the heat needed to melt or vaporise it. Very few self-shielded wires can compete in this respect with their gas-shielded counterparts. Furthermore, some nitrogen pickup by the weld is inevitable and the elements used to control it, such as aluminium and titanium, can have quite undesirable metallurgical side-effects. To achieve satisfactory toughness, for example, it may be necessary to make a weld using a large number of stringer beads rather than in fewer runs by a weaving technique which would involve less inter-run cleaning. The unbiased fabricator, while applauding the ingenuity which produced such a broad and versatile range of products, may regretfully conclude that economics dictate the use of conventional gas-shielded wires.

### Alloyed wires

In principle, most of the wire types described can be made as carbon-manganese or low-alloy versions. The high hydrogen levels and poor toughness which make low-alloy rutile MMA electrodes an unattractive proposition do not apply to the same extent with tubular wires, so nickel-alloyed products capable of meeting offshore requirements are to be found in rutile, metal-cored and self-shielded as well as basic forms.

More highly alloyed wires, especially those for welding steels with yield strengths above $550\,N/mm^2$ and for creep-resisting steels, have generally been made in basic form to maximise the weld toughness and minimise weld metal hydrogen levels. Recently, however, production routes capable of giving hydrogen levels below $5\,ml/100\,g$ even with rutile slag systems have allowed some manufacturers to introduce or reintroduce 1 CrMo, 2 CrMo and high strength ($700\,N/mm^2$ YS) weld metals in rutile form and these will certainly prove serious contenders if claims for them are borne out.

As alloy contents rise, the requirements of the slag system change. Austenitic stainless steels are not susceptible to hydrogen cracking and do not undergo sudden ductile-brittle transitions, so basic slag systems offer few advantages while rutile slags can be used without penalty to optimise the weldability. The lower melting point of the steel compared with low alloy systems makes it rather more difficult to formulate wires which are good for both downhand and positional welding. Downhand types have slags which are too fluid to make small fillets in the vertical-up direction, though some may be used vertically down in smaller sizes. Wires designed for vertical welding have quite stiff slags and can give gas trails on the weld surface in downhand fillet welding unless the maximum current is restricted. Many of the rutile stainless wires available today are of Japanese origin and were designed to run under $CO_2$. They are nevertheless capable of producing weld metals with less than 0.04% C; indeed, the difference in deposit carbon content when the wires are run under mixed gas or under $CO_2$ is usually less than 0.005%.

Because nitrogen is relatively soluble in austenite, and indeed in high chromium steels whether austenitic or not, it is not difficult to make self-shielded tubular wires in the stainless grades. They were popular in the past and are still widely used for surfacing applications. In recent years, however, the weldability of the gas-shielded stainless flux-cored wires has become so good and spatter levels have fallen so low that users now mostly prefer these. Metal-cored stainless wires are also available. They produce somewhat tougher welds than the rutile wires and may be used positionally with dip transfer or pulsed arc. The absence of a slag means that some oxidation of the weld surface occurs as it cools and the welds therefore appear less shiny than those made with flux-cored wires.

A wide range of tubular wires for surfacing is available and some are sold in large tonnages. Surfacing wires are made for gas-shielded, self-shielded and submerged arc welding. As the alloying level increases, it becomes more difficult to make a wire with the attractive weldability of modern low-alloy joining wires: in particular, the slag detachability tends to suffer. However, there is enough variation between the products of different suppliers to make it worthwhile trying more than one. Theory suggests that a quite small number of alloy types should be sufficient to deal adequately with the great majority of applications and experience shows that these only need a few standard slag systems to make them work. Proliferation of products has not been discouraged by manufacturers and partly arises through the willingness of users to believe claims for the miraculous properties of proprietary compositions. Often, a user wanting to change supplier will specify that the product must be of identical composition to that previously used, when a standard product would have performed equally well at lower cost. Users should certainly beware of products offered with no specification, which may well be more expensive than widely used 'generic' products.

## Sheath materials

The easiest type of sheath material to form into a tube is a soft mild steel. Using a more highly alloyed steel has two disadvantages: it is more expensive to buy and, since it undergoes more work hardening, considerably more difficult to process. Users of low-alloy MMA electrodes have sometimes insisted that the alloying additions are made through the core rod for the sake of reliability, though this is now less common in the low-alloy types. Flux-cored wires have escaped this demand, both on the basis that powder cannot escape from a closed tube in the way that coatings can be chipped off electrodes and because the efficiencies with which elements are transferred from the core of tubular wires into the weld metal are very consistent and can be precisely determined. Manufacturers can control filling percentages accurately enough for weld compositions to be met reliably and certification of analysis is usually available.

When stainless and highly alloyed surfacing wires were made in diameters of 2.4 mm or more, it was possible to use high filling ratios and mild steel sheaths. Smaller wires proved more difficult to produce with the thin walls needed so the tube material was made from stainless steel, first with the 17% chromium ferritic type and increasingly using 18/8 steel. Users will hardly be aware of the differences but if the same formulation is made up with ferritic and austenitic strip the latter gives a somewhat higher deposition rate because of its higher electrical resistivity.

## Production methods

Most of the production methods ever used to make tubular welding wire are still in use somewhere in the world. In the 1920s in Europe, steel billets were drilled and filled with fluxing material

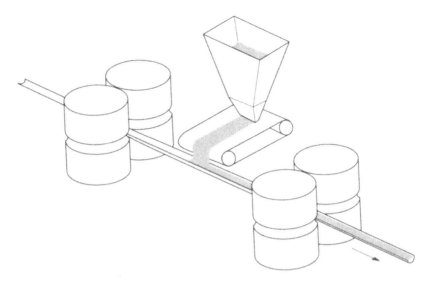

**2.4** Forming and filling tubular wire with a belt powder feeder.

before being forged and rolled down to size. This means of production is still used in Japan, where a new patent application recently described it.[14]

More commonly, wire is produced by forming a thin strip into a U-shape by means of rolls, filling it with powder and continuing to roll it into a circular section, Fig. 2.4. Reduction to final diameter may be by further rolling or by drawing through dies. To feed the powder into the strip, a variety of devices may be used, falling into two broad categories: those which fill the tube to the brim, so that the filling ratio depends only on the geometry of the setup and the density of the powder, and those which meter the powder into the tube so that a range of filling ratios can be achieved. Of the latter, powder feeders based on a continuous belt are probably the commonest. For lines designed to run at high speed, sophisticated design is needed to ensure constancy of fill: among other factors to be considered, the apparent density of the powder may change as it moves faster. The powder has to accelerate rapidly to the speed of the tube and then be prevented from moving longitudinally before it is compacted, to avoid fill variations along the length of the tube. Modern production equipment can perform this task effectively at very high speeds.

When flux-cored wire production began in the UK in 1956, it

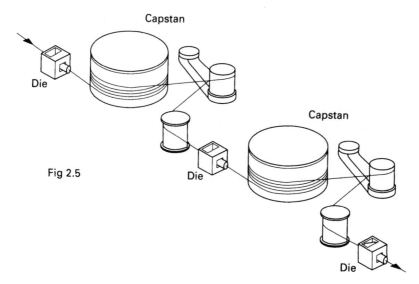

**2.5** Reduction of tubular wire by drawing through dies.

was reduced to size by continuing to process the wire through a series of driven rolls. By this means, the tension applied to the wire is minimised and wires can be made with thin walls and hence high filling ratios. Good roll design also allows the use of sheath materials such as austenitic stainless steel whose work-hardening properties make them difficult to reduce. Only liquid lubricants are needed for the rolling route, and these could be chosen for their volatility so that the heat generated by the production process evaporated them almost completely. By 1968,[23] a 2.4 mm rutile wire made by this route gave a weld metal diffusible hydrogen content of less than 10 ml/100 g at 400 A, a creditable figure even by modern standards. Because it is not easy to control the final wire diameter accurately from rolls, a final pass through a calibration die is now usually part of the production route.

Later other manufacturers reduced the wire diameter with a series of dies and capstans as is normal for solid wire, Fig. 2.5. Whereas solid wire may be drawn submerged under liquid, the danger of liquid penetrating the tube normally precludes this for flux-cored wire. Instead, solid soaps are most often used though it is possible to draw at more modest speeds without these, by applying small, carefully controlled amounts of liquid lubricant. In fact, although soaps may be solid when put into the die box, they become liquid under the high temperatures generated in the die and there is a risk of their being forced through the seam by the high pressure. Even when this does not happen, a residue of soap remains on the wire surface. For these reasons, wire made using solid drawing soaps is usually baked to burn off any hydrogenous residues before it is packed.

Wire which has been properly manufactured should be quite stiff, which contributes to good feeding through the welding torch. If soaps are carefully chosen, their inorganic residues, together with the oxide film left after the wire is baked, may actually improve arc stability and form an integral part of the wire formulation.

To prevent drawing lubricants from entering the tube, and to avoid moisture pickup by hygroscopic flux constituents during storage, wire manufacturers have long sought to make a seamless tube and some have succeeded. The method used for the production of most seamless wire has involved filling a pre-formed and welded tube.[15] This tube, typically about 12 mm in diameter, is rolled into spiral lengths of about 15 m and attached to a vibrating table to encourage the powder to move down it. To ensure that all the components of the filling arrive at the end of the tube at the same time and in the correct proportions, the powder must be agglomerated with waterglass, much like a submerged arc flux. Sections of filled tube are then welded together for reduction to the required diameter. Because the degree of reduction is large, an intermediate annealing process is generally needed and this means that constituents that would react together at the temperatures involved, for example strong deoxidants and weakly bound oxides, cannot be used.

Less commercially successful methods of producing seamless flux-cored wire have welded the seam on-line with the core material *in situ*. The difficulty here arises from the limiting speed of welding, which is usually lower than the speed at which a flux-cored wire line would otherwise operate. To achieve an economic tonnage throughput at a lower speed, the initial tube diameter and wall thickness must be increased, which in turn leads to a requirement to anneal the wire if it is to reach the smallest diameters. In addition, because the tube undergoes considerable reduction before the powder is compacted, there is a tendency for the powder to move longitudinally in the tube, resulting in a varying filling ratio and wall thickness.

Seamless tubular wire may be processed just like solid wire, using wet drawing if appropriate, and the wire may even be coppered. Final processing speeds can be high and the wire can be stiff and resistant to deformation by the drive rolls of the welding equipment, so feeding problems are rare. The good surface quality, especially if the wire is coppered, leads to good electrical contact at the torch tip and to low tip wear. The annealing process should eliminate potential hydrogen from the wire and sealing the tube ensures a long shelf life. For these reasons, seamless wire has been deservedly successful in many areas. However, good deoxidation practice lies behind the remarkable improvements in weld metal toughness achieved with welding consumables of all types in the last 30 years, and it is precisely here that the need for annealing so inhibits the formulist of seamless wires. For offshore and similar applications needing weld metals of the highest toughness, seamless wires have yet to make their mark.

The largest single item of cost in manufacturing most flux-cored wires is the steel for the sheath, which is normally bought in strip form. Hot-rolled rod is much cheaper and a process for making tubular wire directly from rod was patented in 1969.[16] To minimise work-hardening effects, the rod is not rolled flat but a narrow groove is first pressed into it and then widened by subsequent rolls. The powder fill is fed into the groove and the rolling is continued, now closing the U into an O section. The process allows wires of less than 2 mm diameter to be produced although more work hardening does occur than when wire is made conventionally from thin strip. The resulting product is hence very stiff, which gives it good feeding properties. The disadvantage of the system is that wires with a bulky filling, such as self-shielded types, cannot be made in very small diameters.

### Wire section types

However wire is made, the finished product is characterised by a certain ratio of filling to total wire weight, typically in the range 12–50%. Some publications have expressed the ratio as that of fill to sheath, but this is not now common.

Since a given type of filling will compress to a more or less constant limiting density, the ratio of fill area to sheath area in the final tube cross-section depends only on the filling ratio and not on the process route used to produce the wire or the initial dimensions of the strip used. During the reduction process, the tube is first reduced in diameter without thinning of the wall. As the powder inside becomes compacted, it acts as a mandrel so that the tube wall is thinned and the elongation increases. The wall thickness of the tube after the powder is fully compacted depends only on the final diameter and the amount of powder put into the tube at the beginning. In practice, thinner strips may be used to make products with thinner walls so that the amount of mechanical working each receives is about the same, but alternatively it is possible to make products with quite a range of wall thicknesses from a single strip type and this strategy may be chosen as a means of minimising inventory costs. Users need to be aware that the percentage fill will determine the operating characteristics of the consumable while the strip size from which it was made will at most have a small effect on its stiffness.

The majority of wires are made by rolling strip into a tube with butted edges, Fig. 2.6, but if the strip is thin in relation to its width, it may be mechanically difficult to achieve a well butted seam. Furthermore, if the finished product is very thin-walled, any slight opening of the seam could lead to an escape of powder. For these reasons, wires designed for high filling ratios, for example 30% and above, are usually made with the tube edges overlapped.

Early flux-cored wires were of large diameter by today's standards and posed the problem of transferring heat from the sheath to the powder at high wire feed speeds. One solution, as discussed earlier, was the use of large amounts of synthetic titanates, in rutile products, to carry the heat through the core. An alternative approach was to fold the sheath, Fig. 2.7, so that some part of it remained within the tube in intimate contact with the powder. The heat would then be generated more uniformly through the tube and melting would be improved. An incidental benefit of folded sections was greater stiffness and improved buckling resistance when large diameter wires were fed through long conduits.

A final triumph of the roll designers was to produce, from a single strip, a pair of concentric tubes which could contain alloying elements in the centre and slag formers in the outer tube,[17] Fig.

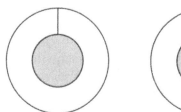 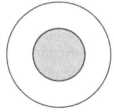 

**2.6** Wire sections, butted (left), seamless (centre) and over-lapped (right).

**2.7** Tubular wires with folded sections.

**2.8** Double concentric folded tube.

2.8. This is used in the former Soviet Union for self-shielded wires and diameters as low as 2 mm are available, but the products do not appear to compete commercially with self-shielded wires from the USA and elsewhere.

### Wire lubrication

If wire has to be fed through a conduit which may be anything up to 25 m long, some form of surface lubrication is needed. When tubular wires were first made by drawing through heavily soaped dies, the soap residue left on the wire was enough to ensure good feeding in use. Now that users demand almost complete removal of any lubricant used in manufacture, other arrangements have to be made. These fall into two broad classes. Solid inorganic lubricants such as graphite and molybdenum disulphide add no hydrogen to the weld metal and are widely used. Their main disadvantage is that they can accumulate in liners and eventually cause clogging of the tip. Alternatively, liquid or solid organic lubricants can be used. These have less or no tendency to clog liners and tips but must be carefully controlled so that any contribution to the hydrogen content of the weld metal is minimised. Products of both types should perform well in most applications if properly made, but in certain cases one or other may be preferred and manufacturers should be able to advise on these.

# CHAPTER 3

# Planning for productivity

The drive to increase productivity has been the main force for welding process change, and tubular wires have been beneficiaries of that change, but planning for productivity should start well before a consumable has been chosen. Early on, it needs to be established whether welding can be done in the flat or horizontal-vertical positions or whether welding in position will be needed. Welding vertically is almost always slower, and may be more so if heat input has to be restricted because of mechanical property requirements. The next stage of planning is the choice of joint type and preparation.

### Joint preparation

In the following discussion, the distinction between welds and joints will be made: *welds* are made as either butts or fillets, Fig. 3.1, whereas the *joint* type describes the configuration of the connected parts, Fig. 3.2. Excellent textbooks[18,19] are available on design for welding in general and only a brief discussion is needed here to relate general principles to the specific requirements and potentialities of tubular wires.

Welding will be simplified if fillet welds can be used instead of butts. No preparation of the joint edges is then necessary except to ensure adequate fit-up of the components, and welding currents are not restricted by the danger of burning through the root. This means that rutile flux-cored wires can be used in all positions provided a suitable type is chosen. If wires formulated for good

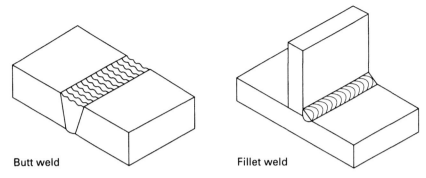

Butt weld                    Fillet weld

**3.1** Butt and fillet weld configurations.

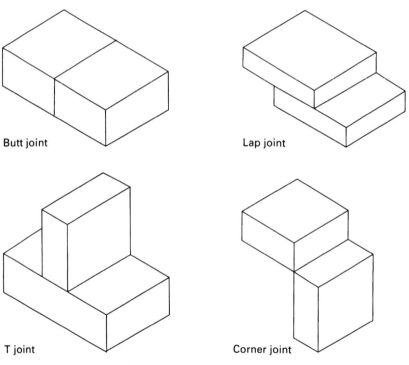

Butt joint                    Lap joint

T joint                    Corner joint

**3.2** Joint configurations.

penetration are used, designers can often reduce the nominal fillet
size in recognition of the improved joint efficiency.

Where butt welds are inevitable, the choice of preparation and
technique for making the root is often critical in terms of both

productivity and quality. The conventional open root with a root face of 1 or 2 mm and a root gap of 2 or 3 mm, widely seen in MMA welding, is satisfactory where basic or metal-cored wires are to be used in dip transfer, but the higher penetration of rutile wires, which always operate in a quasi-spray transfer mode, means that an unrealistic degree of concentration would be needed to make such a root without burning away the edges. Rutile wires are sometimes used with no root gap where a double-sided joint is to be back-gouged before the second side is welded. In that case, the depth and radius of the gouge must be precisely specified and controlled.

Rutile wires are particularly suited for use with various types of temporary backing, especially ceramic backing strips, Fig. 3.3. These may be used for welding either horizontally or vertically with appropriate wires, and give an excellent profile to the back of the joint. In vertical welding, the best profile is obtained using wires with slag systems which are not the very fastest freezing: manufacturers can advise on their most suitable type. Tapes and glass rods have also been successfully employed as backing material and even offcuts of plate glass, but backing products containing borax glass can cause weld metal cracking and should be avoided.

It has been noted that the distribution of heat between the wire and the parent material is largely affected by the amount of resistive heating in the electrical stickout, so if this could be varied, it should be possible to control the root penetration when welding difficult joints with varying root gaps. With gas-shielded processes, changing the stickout without changing the gas shroud soon leads to loss of shielding, so this is not practicable, but with some self-shielded wires quite large variations in stickout can be tolerated. These wires can cope with irregular joints as well as the best stick electrodes. However, the generally lower productivity of self-shielded wires means that this is a second best option compared with achieving accurate fit-up from the start. The temptation to try to get the best of both worlds by mixing self-shielded and rutile wires in the same joint should be resisted since the strong deoxidants in the self-shielded weld metal, diluted into the rutile weld metal, reduce $TiO_2$ to $TiO$ at the slag-metal interface and make the slag almost impossible to remove.

Joint included angles when welding with tubular wire can often

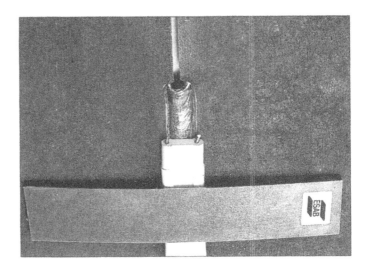

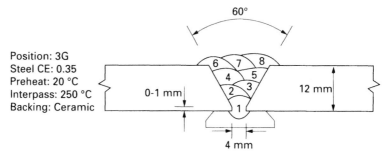

Position: 3G
Steel CE: 0.35
Preheat: 20 °C
Interpass: 250 °C
Backing: Ceramic

| Pass No. | Wire diameter, mm | Amps | Volts | Speed, cm/min | Shielding gas |
|----------|-------------------|------|-------|---------------|---------------|
| 1 | 1.2 | 140 | 21 | 6 | Ar/20% $CO_2$ |
| 2-8 | 1.2 | 180 | 24 | 21-27 | Ar/20% $CO_2$ |

**3.3** Use of temporary backing in welding with flux-cored wire.

be less than with MMA electrodes and this represents a significant productivity advantage, Fig. 3.4. Where a 60° included angle would be normal for MMA, tubular wires can generally use a 50° angle in downhand welding and 40° or less for metal-cored wires.

In vertical welding only basic wires should need 60°, rutile wires 50° and metal-cored wires again 40°, access for the torch being one of the limiting factors here. When making flat butt joints with metal-cored wire it is possible to produce defect-free

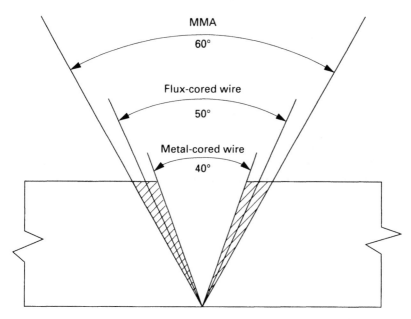

**3.4** Reduction of joint angle when using tubular wire.

welds in parallel-sided preparations no more than 10 mm wide, but the beads will then be directly above one another and toughness will not be optimised. Because the welding torch needs no gas shroud when self-shielded wires are used, access is much better in narrow joints and if a type giving good penetration is used, 40–50° included angles may be possible.

### Wire and gas selection

Unless the demands on the mechanical properties of the joint are unusually severe, basic wires are rarely the first choice of the welding engineer. In practice, yield strengths above about 600 N/mm² or alloying with chromium and molybdenum for creep resistance may call for basic slag systems both to reduce the risk of cracking in weld metal and heat-affected zone and to minimise any deterioration in toughness on stress relief. The more fluid slag of basic wires also reduces defects in horizontal-vertical (2G) butts, where the fast-freezing slag of the E71T-1 types is less helpful than in vertical welding. Basic wires' more fluid slag helps in welding over primer and other forms of surface contamination, though if the contamination can be removed, for example by the use of a gas torch, productivity gains resulting from a change in wire type may make this worthwhile.

Ease and comfort of use will generally point to a rutile wire for most applications, though in downhand welding, the choice between rutile and metal-cored wires may depend on individual preference. Rutile types produce attractive, shiny weld beads and although slag is formed, it is often self-lifting. Heat and radiation levels are lower than with metal-cored wires, especially since wires intended for high current use have traditionally been designed to run with $CO_2$ shielding. In robotic welding, arc radiation is not a problem so the absence of slag and the even lower levels of spatter favour metal-cored wires. Lack of slag also helps metal-cored wires to weld over primers, the relative performance of metal-cored and basic wires depending on the primer formulation. The ability to work with narrower preparations and at high travel speeds has been claimed to improve productivity when a crane manufacturer changed from a rutile to a metal-cored wire.[20]

Only minor differences in deposition rates may be ascribed to the shielding gas, and these are over-ridden by other considerations when it comes to gas selection. The gases most widely used in Europe are argon mixtures with 12–25% $CO_2$. Higher $CO_2$ levels may result in increased spatter and hence weld cleaning time, while lower levels may lead to poorer penetration profiles and an increased risk of lack-of-fusion defects. Users should not be afraid to select a shielding gas primarily to obtain good productivity and freedom from weld defects rather than for metallurgical reasons. Gas manufacturers in particular may give the impression that the gas controls the metallurgy of the weld metal and that a gas with low oxidising potential will allow better mechanical properties to be achieved. While this may be true with solid wires, tubular wires are optimised for particular ranges of shielding gas and those designed to use high argon mixtures may need iron oxide additions to the core to restore the desired oxygen balance. Pure $CO_2$ shielding with wires specially designed for it may allow higher currents to be used without discomfort. Helium-containing gases are generally difficult to justify when welding ferritic steels. In the 1960s, it was discovered that if a very high current density was applied to a solid wire, a stable rotating arc developed and high deposition rates were achieved[21] but penetration was low.

More recently, it has been claimed that penetration could be restored by using an appropriate helium-containing gas[22] and it has been further claimed that the process may be applied with advantage in tubular wire welding. Research institutes are now investigating this. The benefits of helium mixtures are easier to see in welding stainless steels with solid or metal-cored wires for which argon-oxygen mixtures would otherwise be used. In that case some improvement in underbead profile and surface finish may be found.

### Burn-off and deposition rates

The productivity of different types of welding consumable has conventionally been compared by a graph such as Fig. 1.1 or 3.5, showing deposition rates in kg/h. There are cases where this is a valid comparison: for example, in many hardfacing applications where all that is needed is to deposit the greatest possible amount of metal in the shortest time. Unfortunately, the comparison has often been misused, as shown in the previous chapter.

An arc running at a given current and voltage generates a fixed amount of heat which is divided, in welding, between the parent material, the filler metal and losses to the environment. Any change which increases one of these must decrease the others. Where the burn-off rate of the consumable is high, extra care will be needed to ensure adequate penetration. If the burn-off rate is lower, on the other hand, the extra penetration may allow a smaller weld preparation to be used with either a heavier root face or a narrower included angle, so that the joint is completed with less filler. In practice, few welds are made with the power source operating at its maximum rated capacity so if a slightly higher current is called for by a wire with greater penetration, that should not cause any difficulty. As mentioned earlier, welder comfort should not be impaired.

The main area where deposition rates can form a useful basis for comparison between processes and consumables is downhand fillet welding. Rutile, E70T-1 wires in 2.4 mm diameter have routinely been used semi-automatically at up to 600 A, with deposition rates of 11 or 12 kg/h, under $CO_2$. Used in a mechanised application at 700 A with a longer stickout, 19 kg/h has been recorded. More commonly, 500 A would be used with a 2.4 mm

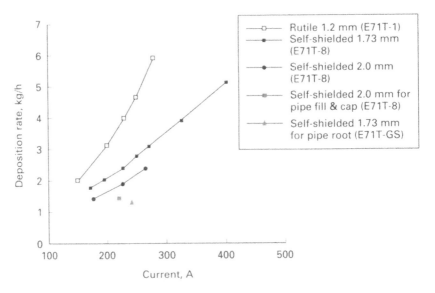

**3.5** Tubular wire deposition rates.

wire to give around 8 kg/h, or 400 A with a 1.6 mm wire for a similar rate but less penetration. Shielding with $CO_2$ greatly reduces the heat of the arc compared with argon-$CO_2$ mixtures and especially helium-containing gases.

In positional welding, it is not the deposition rate per amp which is important but the maximum rate at which the welder can still keep control of the weld. As stovepipe welders know, welding downwards offers the highest productivity since higher currents can be maintained than when welding upwards. Shipyards use metal-cored wires vertically down at 280 A, giving a deposition rate of over 5 kg/h. Fillets are concave but the throat thickness is sufficient for the application. When self-shielded wires have been used in offshore fabrication, vertical-down welding has often been chosen, partly for reasons of productivity and partly because thin stringer beads give better weld metal toughness. In this case, deposition rates are typically 1.5–2 kg/h at 200 A.

For welding vertically upwards, the use of rutile, E71T-1 wires allows currents approaching 250 A and deposition rates of 3.5 kg/h to be achieved semi-automatically and 300 A and 4.3 kg/h in mechanised welding provided no other limitations are imposed by mechanical property requirements. Where heat inputs are likely to be high or difficult to monitor, a 1% Ni wire may be specified for increased confidence in Charpy toughness at $-20\,°C$, even though a carbon-manganese wire would perform adequately under more tightly controlled conditions. In offshore fabrication, many users restrict deposition rates in vertical-up welding to 2.8–3.0 kg/h for optimum weld metal toughness.

Where basic wires have to be used in the vertical-up direction, their more fluid slag means that maximum currents and deposition rates are reduced to around 150–170 A and 2–2.5 kg/h respectively, depending on plate thickness. Furthermore, since the wires are then operating in a dip or globular transfer mode, penetration is less good and a wider joint preparation is needed to avoid trapping slag, so overall productivity falls by proportionately more than the reduction in deposition rate.

Deposition rate data can be helpful in choosing between different welding processes or between different classes of consumable: for example, in suggesting the use of a rutile rather than a basic wire or gas-shielded rather than gasless. However, because of the other factors affecting productivity, such as weld cleaning time, repair time, frequency of tip changes and so on, the small differences which can be measured between consumables of the same class rarely emerge as significant in the final equation. Choices between products of different manufacturers will continue to be made on the basis of user preference, quality and properties of the deposit.

## Welding speed

In welding heavy sections, the use of high travel speeds means lower heat inputs, more runs and more deslagging time. Furthermore, the lower heat input may require a higher level of expensive preheat. Only in offshore fabrication with self-shielded wires, where vertical-down stringer beads made at high speed are needed in any case to achieve the required toughness, has this been a profitable strategy.

In single pass welding, on the other hand, the aim is to maximise travel speeds while meeting the geometrical specification of the joint. In shipbuilding, where structures have a high degree of redundancy, fillet weld sizes are often quite small, for example 6 mm leg length or 4 mm throat thickness, so vertical-down welding at high speed can be the most productive technique.

For welding horizontally, the limiting speed is not likely to be set by the process if welding is semi-automatic: welders will not be comfortable at travel speeds above 0.5 m/min and most manual welding is slower than this. Where welding is mechanised, tubular wires can operate at speeds above 1 m/min. On clean plate, single-run butts at 1.3 m/min are feasible with rutile and metal-cored wires. Speeds up to 3 m/min have been achieved in circumferential welds in 300 mm diameter, 1.6 mm thick cylinders using a joggle joint, a 1.8 mm metal-cored wire at 450 A and the welding head at the 2 o'clock position.

The effect of contamination or paint on a surface to be welded in causing porosity increases with welding speed because the faster-freezing weld pool traps evolving gas. Basic and metal-cored wires are least affected by extraneous contaminants, but even they have to be run at reduced speed over primers. In shipbuilding especially, this is an important factor determining productivity. There is no universal recommendation for the best type of wire for welding over primer, but since welding manufacturers and paint manufacturers now work closely together, advice on optimum proprietary combinations should be available from them.

When welding speed in a straight line is the over-riding criterion in consumable selection, only specially designed, self-shielded tubular wires can at present match the 5 m/min achievable with the submerged process under ideal conditions. To do this the joint is inclined at an angle of up to 20° to the horizontal with the direction of travel downhill.

## Pulsed arc welding

Pulsed arc welding was developed in the 1960s and achieved some commercial success in the 1970s for aluminium alloys. With the increasing popularity of inverter-based power sources in the 1980s, allowing a pulsing facility to be provided at little or no extra cost, many users found themselves with access to pulsed power, but for steel welding the process still looked to many like a solution in search of a problem. Over-enthusiastic promotion of the process has also obscured some of its real benefits.

The original aim of pulsing the arc was to maintain free-flight metal transfer at lower mean currents than would be necessary using conventional power. For example, a wire with a spray transition current of 300 A can be used with a pulse above this level but a mean current of 150 A or less. In principle, this allows positional welding in the free-flight mode, with consequently improved penetration. A second benefit came when the move from thyristor to inverter technology permitted stepless changes in pulse frequency at reasonable cost. By varying the frequency in proportion to the wire feed speed, each pulse can be made to relate to a fixed length of wire: if the constant of proportionality and pulse parameters are correctly chosen, each pulse melts and detaches a single droplet whatever the wire feed speed and mean current. Thus a wide range of welding currents can be called up using a single knob and with no discontinuity in metal transfer characteristics. The term 'synergic pulsed welding' was coined to describe this mode of control.

This latter feature could have been the biggest selling point for pulsed arc welding but paradoxically, part of the technology package that made it possible also helped to make it less necessary. When once the arc is controlled electronically, it is an easy matter to include a microprocessor in the system to permit 'one-knob control' in a non-pulsed mode however complex the voltage-wire feed speed relationship. Many modern power sources feature pre-programmed or mapped settings which, when the user has selected the wire type, automatically optimise the arc voltage and even an analogue of the inductance in response to user-driven changes in wire feed speed/current. Many tubular wires benefit from the improved ease of setting-up and use which this affords though productivity is not necessarily directly affected.

One of the more significant features of pulsed arc welding was for many years ignored because the research organisation developing the process proposed that the wire burn-off rate was proportional to the mean current. In fact, because much of the heat used to melt the wire is developed in the stickout as $I^2R$ heating, pulsing the arc increases the burn-off rate per amp of mean current, the more so when short, high current pulses are used so that the mean square current moves away from the square of the mean current. In terms of deposition rate, pulsed arc welding produces the same effect as increasing the stickout, but without the associated problems of maintaining the gas shield. In a shipyard application making HV (2F) fillets, a basic wire was used to give good performance over primer and pulsed welding to raise the productivity to levels typical of rutile wires. The ability to transfer 'one drop per pulse' also reduced spatter.

For positional welding, increasing the wire melting rate at the expense of heating the parent plate may lead to poorer weld pool control and a risk of lack of penetration, so pulsed arc welding is not an automatic choice. However if, for example with basic wires, the alternative is dip transfer, with potentially poor penetration and high spatter, a good pulsed system may help. Control algorithms for pulsing basic wires are relatively new and a recent refinement allows the use of variable pulse shape to adjust the heat balance between wire and parent material between horizontal and vertical welding. This is discussed in more detail in Chapter 6, 'Equipment'. Again, the power sources which provide the best control in pulsed mode are likely to offer some form of enhanced dip transfer, which will be preferred for difficult joints with basic wires.

One area where pulsed welding in the vertical position is able to demonstrate a convincing advantage over continuous current welding is when using metal-cored stainless wires. With a special pulse program developed for the purpose, the welding speed for vertical-up fillets can be increased by 30% while maintaining an excellent mitred bead profile even for small fillets of 4 mm leg length (3 mm throat). Solid wires also benefit from pulsing, but the welding speed achievable falls short of that when using tubular wire.

### Mechanisation and robotics

If a job can be mechanised or carried out by a robot, productivity gains follow for both direct and indirect reasons. First, taking the operator away from the heat and fume of the arc allows less user-friendly conditions and consumables to be used: 900 A with a gas-shielded rutile wire or 500 A with a metal-cored wire designed for welding armour plate are practicable only if the equipment is mechanised. Similarly, any welding speed above 0.5 m/min calls for mechanisation. Even at lower currents and speeds, welder fatigue can limit duty cycles.

A large proportion of robotic welding installations at present use solid wire. Those which use tubular wire pay more for the wire so must justify its use in terms of productivity or quality.

Most use metal-cored wire to minimise deslagging. The main advantage over solid wire, when using argon-rich gas mixtures to reduce spatter, is the improved underbead profile. With no welder to guide the torch along the joint centre, this gives greater confidence in achieving full penetration. In a setup for welding thin material where some distortion was occurring despite the jigging, the welding speed was increased from 1 to 1.3 m/min by changing from a solid to a flux-cored wire. In another robotic application, a 1 mm metal-cored wire replaced a 1.2 mm solid wire, giving a better bead profile, lower defect levels and increased productivity.

Productivity often gains from mechanisation by more than the effect of the process change itself. Where welding is seen as a low technology process, the infrastructure needed to make it work efficiently is often neglected: parts to be welded are not delivered on time, the welder has to go to another part of the site to fetch consumables, equipment is not subject to planned maintenance. When an expensive robot replaces a manual operator, managers and accountants take an interest in the reasons for its down-time and measures are taken to ensure that work flows smoothly past it. Tubular wire, for example, can be delivered in packs of 300 kg from which it emerges straight and kink-free, eliminating 20 reel changes per pack. This has naturally proved popular with robot users but manual welders also represent an expensive resource and benefit similarly from more convenient ways of delivering the consumable.

# CHAPTER 4

# Planning for quality

Any organisation which has costed the repair or reworking of defective welds knows that this can multiply the cost of the original weld several times. Tubular wire welding offers high weld quality with low repair rates provided the origins of certain types of defect common to most welding processes are understood and simple precautions are taken against them.

The most common weld defects, or as British Standards now refer to them imperfections, are lack of fusion, porosity and hot or cold cracking. In addition, the geometry of the joint or its mechanical properties may not meet the specification. This chapter will discuss ways of ensuring that the weld is right first time and will conclude with some hints on troubleshooting.

## Lack of fusion

The danger of failing to achieve proper fusion has been the main factor in limiting the growth of the solid wire MIG process into high integrity applications previously welded with MMA electrodes. In MMA welding, very little resistive heating takes place in the core of the electrode so most of the electrical energy provided by the power source is available at the arc, while the ceramic cup formed at the electrode tip directs the arc into the joint and, in the case of heavily coated electrodes, cuts down much radiant heat loss. By contrast, MIG welding relies to some extent on resistive heating of the wire so a lower proportion of the total energy is available to melt the parent metal, while the arc can be relatively undirected, for example in the dip transfer mode.

Lack of fusion is thus a problem against which users of MIG welding have to be permanently on their guard, especially since the difficulty of detection of this form of defect makes it a particular *bête noire* of inspection authorities.

The successful completion of several gas transmission lines in South Wales, East Anglia and the USA in the 1960s showed that even with dip transfer, solid wire MIG could be used in pressure systems. These lines were welded with $CO_2$ and benefited from the better penetration profile which that gives compared with argon-rich mixtures: some also used special wires designed to give 30% lower burn-off rates and consequently better fusion than standard wires. Tubular wires, even those types which require the dip transfer mode for positional welding, achieve good penetration profiles and a lower defect incidence by affording more control of the arc and burn-off characteristics and of the wetting properties of the molten metal.

In avoiding lack-of-fusion defects, the first essential is to ensure an appropriate joint preparation as described in Chapter 3. When using basic or metal-cored wires to root in the dip transfer mode, the root gap should ideally be 3 mm with a small root face of no more than 2 mm. Joint included angles should be at least 60° for basic wire and 40° for metal-cored wire, though on thicker (>40 mm) plate a compound preparation is acceptable.

Welding should be carried out with as short a stickout as practicable. Some manufacturers offer alternative lengths of gas shroud to facilitate this, so that for vertical welding the contact tip is flush with or proud of the end of the shroud while for downhand welding it can be set back within the shroud by a few millimetres so as to allow a longer electrical stickout while preserving good shielding, Fig. 4.1. With self-shielded wires, as described earlier, varying the stickout is a legitimate means of coping with variations in fit-up, but those types designed for positional welding, especially of pipes, can deliver such good penetration that fusion defects can be avoided even in multi-pass welds made vertically downwards.

Currents should be maintained within the manufacturer's recommended range and that covered by the procedure approval. If continuous monitoring of wire feed speed (wfs) and current is employed, an increase over time in the ratio wfs/current indicates that the stickout is increasing and that penetration may therefore be at risk.

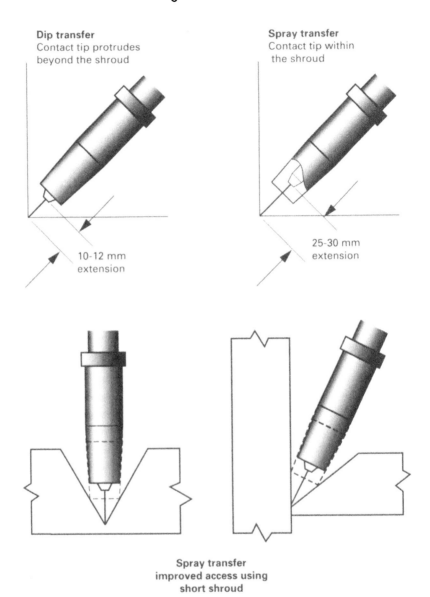

**Dip transfer**
Contact tip protrudes
beyond the shroud

10-12 mm
extension

**Spray transfer**
Contact tip within
the shroud

25-30 mm
extension

Spray transfer
improved access using
short shroud

**4.1** Use of different shroud lengths in gas-shielded welding.

### Porosity

Porosity is one of the commonest defects in arc welding and in the early days of flux-cored wire it was particularly prevalent. Despite this, weld failures attributable to porosity are rare, since pores are not effective stress concentrators. However, now the causes of porosity are better understood, it can be avoided with quite simple measures.

Most gases are more soluble in molten steel than in the solid, so if appreciable amounts of dissolved gas are present in weld metal at the time of solidification there is a risk of porosity. The pores may be spherical or elongated, the latter type sometimes being referred to as 'worm holes'. Related phenomena are seen when gas escapes from the weld metal itself but is trapped beneath a viscous slag, or when nucleation of gas bubbles takes place at the slag-metal interface. In these cases, evidence of gassing is seen on the weld surface when the slag is removed, Fig. 4.2. This may take various forms from small indentations which are difficult to see, through larger indentations or 'gas flats' to elongated tracks or 'worm trails' which may extend for many millimetres.

Three gases or their derivatives cause porosity in welding. Oxygen, either from the atmosphere when a gas shield is lost or from heavy oxides on the surface to be welded, gives rise to carbon monoxide (CO) bubbles in the weld. Nitrogen, almost invariably from atmospheric contamination, forms nitrogen bubbles. Hydrogen comes in the form of water, hydrocarbons and a variety of organic contaminants in the joint preparation or, if some accident has befallen it, on or in the wire itself.

Although many users instinctively feel that more deoxidation would reduce porosity, modern tubular wires are mostly more strongly deoxidised than their predecessors of the 1960s and early 1970s and are not especially susceptible to oxygen-induced porosity. Where there is ingress of air through lack of shielding, it is the nitrogen that will usually be responsible for any porosity that occurs: weld metal oxygen contents can exceed 1000 ppm without porosity and many consumables contain iron oxides as a deliberate addition. The demise, in relative terms, of the AWS E70T-2 wires, originally more strongly deoxidised types capable among other things of welding over scale better than the E70T-1 wires, parallels the eclipse of the 'triple-deoxidised' solid MIG

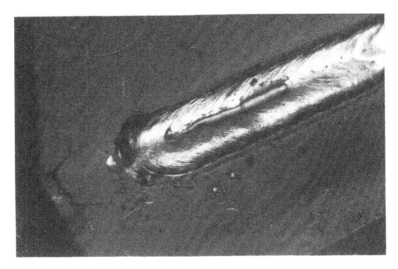

**4.2** Gas trails on the surface of a weld made with rutile flux-cored wire.

wires which once claimed nearly half the UK market. Many of the E70T-2 wires relied for deoxidation on manganese, a weak deoxidant but a powerful strengthening agent, whereas modern E70T-1 wires use a balanced mix of deoxidants which do the same job without excessive increases in strength or loss of toughness. For welding over badly scaled plate, users should find these wires give little away to the old T-2 types. In extreme cases, especially when welding over primer, the basic T-5 wires may be needed.

Nitrogen is an insidious contaminant of welds since its source, unlike rust or grease, is invisible. Unless steps are taken to deal chemically with nitrogen, porosity will typically set in when the weld nitrogen level reaches 300–400 ppm. Unfortunately, the strong nitride-formers needed to react with this level of nitrogen have unwelcome side-effects which so far have limited their use to self-shielded formulations. Gas-shielded welding is therefore relatively unprotected against temporary breakdown of the shield.

Atmospheric entrainment is prevented in the first instance by proper maintenance of equipment and proper choice of gas flow rates. Damaged connectors and leaky hoses not only reduce the amount of gas delivered to the torch, but allow that gas to be contaminated. It may seem at first sight that a positive flow of gas outwards from a leaking tube should prevent an inward flow of

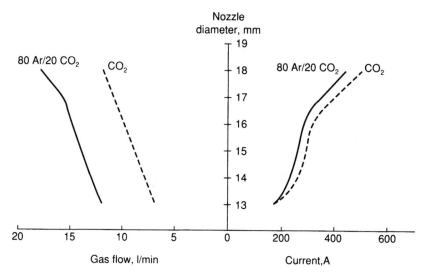

**4.3** Nomogram for selecting nozzle diameter and gas flow rate for different gases.

contaminants, but if the partial pressure of nitrogen inside is close to zero, nitrogen will diffuse inwards while the shielding gas escapes. Before starting to weld, adequate purging time should be allowed, especially of long hose systems. Consideration may need to be given to back-purging the root in sensitive systems, though this is generally needed to combat metallurgical effects of nitrogen contamination rather than porosity.

Gas flow rates should be chosen to suit the nozzle diameter and welding current, Fig. 4.3. While too little gas allows air to enter the arc area, excessive flow rates cause turbulence and shielding again breaks down. Lighter gases need higher flow rates: 30% helium mixtures, used with metal-cored stainless wires, require 1.5–2 times the flow rate of argon-based mixtures. Stainless wires are however not very sensitive to nitrogen-induced porosity because both the high chromium content and the metallurgical structure raise nitrogen solubility in solid steel: good shielding is important here mainly to preserve a clean bead appearance.

Gas shields are easily disturbed by draughts, and gas-shielded welding out of doors needs particular care, although numerous cross-country pipelines attest that it can be done with suitable protection. Gas flow rates have been increased by up to 100% in field pipe welding: a modern mechanised system using a rutile flux-cored wire is operated with 40 l/min of Ar-20%$CO_2$. In the welding shop, open doors on different sides of the shop can cause problems and welding sets should be positioned so that the exhaust from cooling fans does not impinge on the welding area. Fume extraction equipment should be placed so that it does not suck away the gas shield.

Self-shielded wires are provided with protection to ensure freedom from nitrogen porosity in normal use, even out of doors or in draughty conditions, and its appearance is generally a sign that too short a stickout or too high a welding voltage has been used. However, because there is a positive interaction between nitrogen and hydrogen pickup by the weld, moisture can also be a source of problems if the weld preparation is damp or the wire has not been properly stored.

Hydrogen and its compounds are probably the commonest cause of porosity in flux-cored arc welding. In the 1960s and 1970s, much of the hydrogen came from the wire itself. Core materials contained hygroscopic powders, and although the powder filling may have been baked before use, moisture pickup was rapid once the wire was unpacked. Traces of drawing soap were left on the wire. Diffusible hydrogen levels above 30 ml/100 g were recorded.[23]

Modern formulations and production methods have reduced the sources of hydrogen in and on the wire to the point where weld porosity is a sign that the wire has been abused in some way. Manufacturers' packs are generally well enough sealed but out of the pack, some wires can still absorb moisture from damp atmospheres. Even wires with low hydrogen potential can occasionally produce surface gas flats under argon-rich gases at high currents and with short stickouts. A small increase in stickout should be enough to eliminate the problem.

Special techniques are needed to weld over primers, where again the evolution of hydrogen from organic components is usually the main cause of porosity. More oxidising gases are therefore helpful: $CO_2$ is better than Ar-$CO_2$ and Ar-$CO_2$-$O_2$ may be best of all. Slower welding speeds allow more time for harmful gases to escape. When making double-sided fillet welds, it may be helpful to leave gaps in the weld on the first side through which gases evolved during welding of the second side can escape: the gaps are filled when the second side has cooled. More detailed advice on welding over primers can be obtained from the larger welding consumable and paint manufacturers, who work closely together on the technique.

Although sulphur is normally of concern only as a potential cause of cracking, in the presence of moisture it can be a source of weld porosity through the formation of $H_2S$. This would only be expected at sulphur levels found, for example, in free-cutting steels and a change to basic consumables should solve the problem.

In welding pipelines, the root pass is sometimes made with cellulosic electrodes even when semi-automatic or mechanised processes are to be used for filling and capping. Such electrodes can give up to 100 ml hydrogen per 100 g deposited metal, and pipe welding procedures designed for them take this into account. If the hot pass is made with a rutile tubular wire, hydrogen from the root may cause surface scarring on the hot pass surface. Careful selection of the wire and fine tuning of the welding parameters should avoid this.

### Hot cracking

Hot cracking, more specifically solidification cracking of welds, preoccupied welding engineers and consumable manufacturers in the 1960s but like the diseases of the slums has been largely defeated by hygiene. Solidification cracking happens when shrinkage stresses in the cooling weld act on intergranular films of impurity-rich, low-melting-point phases. The two most potent impurities are sulphur and phosphorus and their effect is enhanced by carbon, which promotes the formation of austenite in which they are less soluble than in ferrite.[24] Since the 1960s, sulphur, phosphorus and carbon levels have all fallen steadily in the so-called 'weldable' steels while even in high carbon steels not originally intended for welding, impurity levels are low enough to make solidification cracking a rarity.

If solidification cracking does occur, it can be recognised because it is found immediately the weld is complete, with no incubation period as for hydrogen cracks. The cracking itself usually follows the weld centreline and on breaking open a crack, its internal surfaces are generally blued. Under the microscope, the cracking follows columnar solidification boundaries and there may be significant relative displacement of the crack surfaces. Electron microscopy provides conclusive evidence of solidification cracking, revealing rounded dendrite tips, sometimes with toffee-like spikes where the surfaces have pulled apart, and at higher magnification often films of impurity phases and thermal striations.

Practical prevention of hot cracks depends primarily on ensuring the correct weld shape. In particular, the depth-to-width ratio should not exceed about 1.2.[25] Higher values usually occur through inappropriate joint preparations and cause increased segregation of impurities at the weld centreline. Very wide, flat beads can also be susceptible to solidification cracking because any incipient cracks are rapidly frozen in without any opportunity for 'healing' by the remaining molten metal. If the weld profile is good, only the most severe contamination of the weld will cause cracking. Weathering steels containing phosphorus and copper, painted surfaces containing sulphur or phosphorus and even high sulphur free-cutting steels can all be welded with suitable procedures. In all cases where a risk of hot cracking can be identified in advance, slower welding speeds are helpful. Where sulphur is the potential cracking agent, useful desulphurisation can be achieved by using basic consumables, but no dephosphorisation is normally possible in welding.

### Cold cracking

Cold cracking is always associated in practice with hydrogen embrittlement and may be found in either the weld metal or the heat-affected zone (HAZ). Procedures for avoiding the latter are set out in BS 5135 and the forthcoming European standard EN 1011. Few countries outside Europe have developed consistent systems for measuring hydrogen and using the results to write welding procedures that will avoid cracking: in many cases the hydrogen measurement is seen as an end in itself and the user is left with a mere number to make of it what he (or she) will. The European approach is based implicitly on controlling the HAZ hardness to a level at which the hydrogen level from the consumable will not give rise to cracking: the lower the hydrogen, the higher the permissible hardness. To achieve this, nomograms involving combined plate thickness and weld heat input, together defining potential cooling rate, and plate carbon equivalent, defining hardenability, are provided for each specified hydrogen level or critical hardness to allow the user to work out the required preheat. The user selects the appropriate nomogram on the basis of information on expected hydrogen levels obtained from the consumable manufacturer. He can then either calculate a suitable preheat given a set of welding parameters which he has already decided upon, or fix the preheat and find from the standard what range of heat inputs will be acceptable. This way of working out welding procedures has been in use in the UK since the first publication of BS 5135 in 1974 and its great success has been not only in preventing HAZ hydrogen cracking but in the control of costs through the avoidance of unnecessarily high levels of preheat. From the point of view of the tubular wire user, the only question is how to ensure that he obtains the most meaningful hydrogen level on which to base the calculation.

MMA electrodes are characterised by a single figure denoting hydrogen content. For any given electrode, only one size is normally required to be tested and the current is fixed, in BS 6693 Pt2 for example at 15% less than the manufacturer's specified maximum. MMA welding is a simple process where the range of currents used with a particular electrode size is quite small, so that little change in metal transfer mode occurs. Furthermore, there is no analogue of the stickout variation possible in semi-automatic welding. If any small changes in weld hydrogen level do arise with changes in current or electrode size, they are allowed for to a first approximation because the same variables will have been used in the experiments used to produce the original nomograms.

There is not much experimental evidence on the validity of using a single hydrogen value for a flux-cored wire. For rutile wires in particular, weld hydrogen levels increase with current and decrease with stickout by factors which may be as high as 2 or 3 over the normal operating range of one size of wire, Fig. 4.4. There is also a tendency for hydrogen levels to increase with wire diameter because of reduced $I^2R$ heating in the stickout and to decrease with increasing oxidation potential of the shielding gas, Fig. 4.5.

When the British Standard for flux-cored wires, BS 7084, was

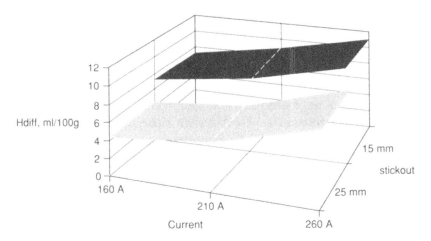

**4.4** Variation of deposited metal hydrogen level with current and stickout for a rutile flux-cored wire.

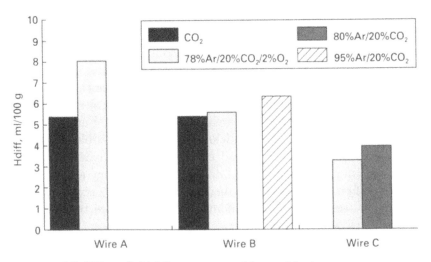

**4.5** Effect of shielding gas on weld metal hydrogen content.

written it was decided to fix the minimum current used to measure hydrogen for classification purposes for each wire size. Manufacturers may opt to use higher currents, but may only use lower currents where their own recommended maxima are lower than those specified. They are also obliged to define ranges of voltage, current and shielding gas for which the claimed hydrogen levels are not exceeded. Some users have asked for supplementary data for the higher currents they are using in fabrication, but it is not thought that there are any recorded cases of HAZ cracks arising through the use of hydrogen values derived according to BS 7084. When this is superseded by BS EN 758, no current levels for testing will be laid down: instead, the manufacturer will have to state the envelope of conditions within which the hydrogen level claim is met. This may lead to the use of higher preheats, but more experimental work is needed to establish the degree of conservatism.

Until recently, weld metals were designed with lower hardenability than most parent steels, so if procedures were aimed at preventing HAZ cold cracking, the weld metal was almost certainly protected too. However, steel designers are now able to take advantage of better controlled process routes so that the metallurgy of the base material is now much closer to that of the weld deposit. This means that its inherent susceptibility to hydrogen-induced cracking may not be very different either. Furthermore, as long as the weld metal transforms from austenite to ferrite at a higher temperature than the parent material, then during the time when they have different structures hydrogen is pushed out of the weld into the surrounding austenitic material in which it is more soluble. As the difference in transformation temperatures decreases, so does the driving force for hydrogen expulsion. If the HAZ actually transforms first, which is possible with some very low carbon steels, hydrogen may be rejected back into the weld metal.

These factors have in recent years tended to shift the risk of hydrogen cracking away from the HAZ and towards the weld metal, not just for tubular wire welding but for other processes as well. Unfortunately, the risks have not been quantified as well as they were for the HAZ when, in the 1960s, funds and resources for the extensive testing needed were more freely available than today. The rules for avoiding weld metal cracking are likely to be quite different from those for the HAZ. Lower cooling rates produced by higher heat inputs produce much less beneficial microstructural changes, while the corresponding increase in bead cross-section means the hydrogen has farther to diffuse out so the net change may be an increased risk of cracking. The effect of interpass temperatures and time between runs in multipass welding is complicated and has not yet been expressed in a form useful to working welding engineers. At present, most weld metals remain more tolerant to hydrogen than the steels they are used to weld. New types under development aim to maintain this balance in the future, but there will be an increasing need for test procedures to assess this as new materials appear.

Apart from the variation in diffusible hydrogen values observed when the welding parameters are changed, as discussed above, there is an irreducible statistical variation resulting from the inhomogeneity of welding consumables themselves and the difficulties of the experimental techniques. The expectations of many who set out to codify the welding process have been absurdly high in this respect. Given that the standard unit of diffusible hydrogen, 1 ml/100 g deposited metal, corresponds to 0.90 ppm, a precision of 1 ml/100 g would compare favourably with that achieved in determining other light elements such as boron. Yet decisions are continually being made on the basis that levels of 4.8 and 5.2 ml/100 g separate sheep from goats or require a different course of action on the part of the user. In fact, the standard error of the mean of a triplicate determination is generally around 0.6 ml/100 g, so any figures after the decimal point must be treated with caution.

Similar caution is needed in allowing for batch-to-batch variations in welding consumables. In the days before modern instrumental techniques for hydrogen determination became available, it was common for manufacturers to check hydrogen levels in their consumables once or twice a year and to quote the figures obtained for use in preheat calculations. There is no evidence that this led to cracking or weld failures, but inevitably users did not see the distribution of hydrogen values that batch analysis would have revealed. When every batch of the same product is analysed, statistics predict that in 100 batches, one will exceed the mean by three standard deviations and this may come as a shock to users. On the one hand it has to be recalled that the product itself has not changed, so neither presumably has the risk of cracking. On the other hand, a better appreciation of statistical effects on the part of those who write and those who use specifications and of suppliers and purchasers of consumables will certainly reduce misunderstandings and produce safety and cost benefits.

These caveats may convey an impression that hydrogen measurements are so imprecise, subject to manipulation and uncertain in their application as to be of little use to practical welding engineers. In practice, though, hydrogen cracking in any form is rare at present and the simple application of BS 5135 or prEN 1011 using hydrogen values measured according to BS 7084 will be sufficient protection. Further co-operative work between steelmakers, fabricators, consumable manufacturers and research institutes will however be needed to ensure that future standards keep pace with developments in materials.

## Metallurgical quality of the weld

Even though a weld may be made with no cracks or physical discontinuities, it may have within it metallurgical inhomogeneity or features which could cause embrittlement or weakening. Flux-cored wire is no more vulnerable to these than other types of welding consumable but the early history of the process may have left some welding engineers in need of reassurance.

To deal first with a concern still often voiced, modern wires are made with filling ratios closely controlled so that the spectre of empty or partly-filled tube should have been banished. Various forms of on-line monitoring are used for this, generally under microprocessor control, and manufacturers are able to provide quantitative data on the degree of consistency achieved. Instrumental techniques for chemical analysis, not available when tubular wires were introduced, and electronic data processing also allow more detailed statistical analysis of the finished product. Manufacturers may legitimately be pressed for information on how they aim to ensure consistent filling and how successful they are.

More difficult to detect on line than variations in fill ratio is segregation of individual elements within the powder blend. This can occur when powders of different density or particle size are mixed together: finer or denser materials can percolate downwards through the blend during subsequent handling. Control of this problem is a matter of careful selection of powders in formulation and good housekeeping during manufacture. Again, manufacturers can provide information on the homogeneity of their products.

One other form of inhomogeneity in the weld has been related to the design or manufacture of the consumable. In both tubular wires and MMA electrodes, alloying elements are added to the flux either in elemental form or as ferro-alloys and can sometimes fail to melt or disperse completely under certain conditions. In that case a microstructural island, often crescent-shaped, of higher hardness may remain in the weld after solidification. Consumables do vary somewhat in their susceptibility to this problem, but welding parameters are also a major factor and in tubular wire welding it is usually symptomatic of an arc voltage above the optimum level.

Contamination of the weld by impurities which adversely affect its properties is commoner than is often realised. Sulphur and phosphorus can be picked up from paints and phosphorus from some weather-resistant steels. There have even been cases of air-borne sulphur contamination causing welding problems in densely industrialised environments. In addition to the risk of hot cracking already discussed, sulphur and phosphorus have a strong embrittling effect on weld metals and much of the improvement in steel and weld metal toughness over the past thirty years can be attributed to their reduction. But the most frequent, and often least suspected contaminant giving rise to weld embrittlement is nitrogen. As already pointed out, nitrogen entrainment arising from poorly maintained or wrongly chosen equipment can lead ultimately to weld porosity, but only a tenth of the amount needed for this may impair the weld toughness. Welding wires would typically contain no more than about 50 ppm of nitrogen and no more should be picked up between wire and weld. Welding in a draughty shop can easily double this to give a serious reduction in toughness, Fig. 4.6.

The new draft European Standard for shielding gases, prEN 439, allows up to 3000 ppm of nitrogen in 80% Ar, 20% $CO_2$ mixtures, over four times the level permitted by the existing German standard DIN 30 526. As shown in Fig. 4.7, this could typically introduce 30 ppm of nitrogen into the weld metal. Some offshore fabricators who enforce strict specifications for the amount of nitrogen in the steel strip used to form the sheath of flux-cored wires have used shielding gases capable of adding an equivalent amount. Today, most gas suppliers offer low nitrogen

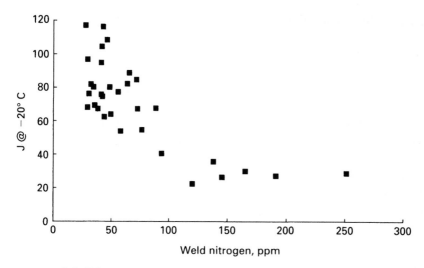

**4.6** Effect of nitrogen on weld metal toughness for a rutile flux-cored wire.

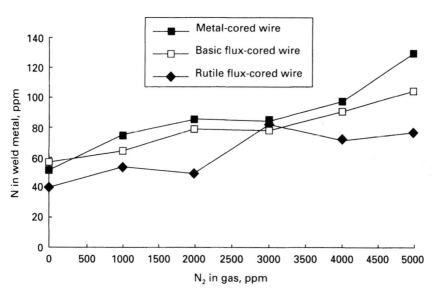

**4.7** Relationship between levels of nitrogen in the shielding gas and in the deposit.

gases either as standard or as a special premium range for critical applications.

## Microstructure and mechanical properties

Flux-cored wires can be used to make welds with mechanical properties as good as those produced by any other process provided they are properly chosen and used. However, because they may be used over a much wider range of welding parameters and heat inputs, care has to be taken that production welding conditions do not stray too far from those of the original procedure test.

Welds rarely fail to achieve adequate tensile properties. While steelmakers have concentrated on moving to leaner compositions through the use of accelerated cooling and thermomechanical processing, weld metals have thermal and mechanical cycles imposed upon them by the joint type and welding parameters, so compositions have remained relatively static. Users have always demanded weld metals which overmatch the parent material in terms of yield and tensile strength and consumable manufacturers have been able to ensure that this happens.

Toughness is unfortunately less simple to control and is more influenced by the vagaries of welding procedures. The commonest cause of poor toughness, and one which affects most arc welding processes, is probably excessive heat input. Good toughness depends on developing a fine grain size in the weld metal and high heat inputs lead to coarser microstructures. In some weld metals, especially self-shielded types, there is a large difference in toughness between the as-deposited regions of a multi-pass weld and those parts whose structure has been refined by the heat of a subsequent pass. In that case, care must be taken to maximise the amount of refined metal by ensuring that the layers of weld metal are thin; a vertical-down procedure is often used. Difficulties can also arise because of the way in which procedure tests sample the weldment.

Tests are sometimes made using a K-preparation so that notches can be placed in the HAZ parallel to the fusion boundary, Fig. 4.8. On the other side of the boundary, weld runs will be stacked one above the other so that a notch in the weld metal could pass

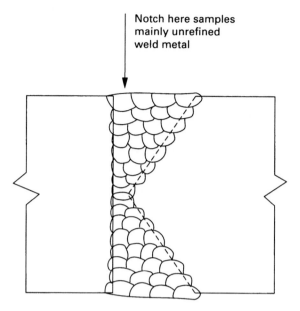

Notch here samples
mainly unrefined
weld metal

**4.8** Sampling weld metal deposited in a K-preparation.

through the centrelines of many runs, sampling an unrealistic amount of coarse-grained microstructure. If the preparation is designed to simulate a real joint, this is reasonable, but if it is designed only for convenience of sampling the HAZ, the weld metal may be unfairly penalised.

Over the last thirty years, the weld metal microstructure which has become most widely used where high toughness is needed is that known as acicular ferrite, Fig. 4.9. The fine grains of this structure promote good strength and toughness at the same time,[26] unlike most metallurgical strengthening mechanisms. Consumable manufacturers normally formulate their products to produce high levels of acicular ferrite in the sort of all-weld-metal test demanded by BS 7084 and the ship classification societies. When welding vertically upwards, welders tend to use a wide weave unless otherwise instructed and this will generate a higher heat input and a less tough microstructure. Attempts to formulate consumables for good performance at high heat inputs can lead to higher tensile strength and hardness when the consumable is used for fast stringer beads and this may be unacceptable. However, the

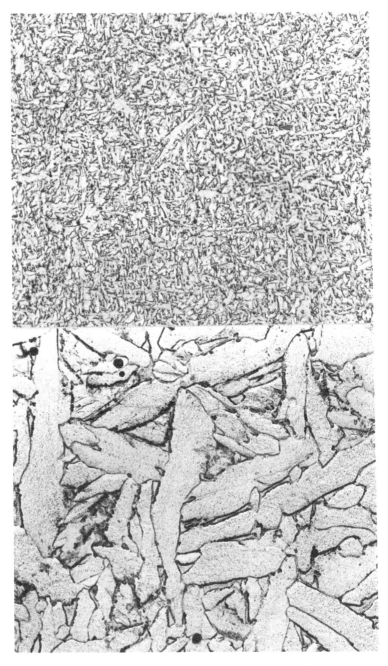

**4.9** Acicular ferrite microstructure, top ×200, below ×4000.

addition of 1% Ni to a carbon-manganese consumable greatly improves its tolerance to high heat inputs without excessive hardening and products of this type are now widely used even when the Charpy toughness requirement is for no lower than $-20\,°C$.

Apart from the heat input chosen, dilution by parent material is the other factor influencing microstructure over which the consumable developer has no control. With self-shielded and gas-shielded flux-cored wire this is rarely a problem unless the parent plate is of particularly exotic composition since the dilution is generally low. In submerged arc welding, on the other hand, where tubular wires are increasingly used, dilutions often rise above 50%. Furthermore, this may be a two-pass process where a high heat input or a thin parent plate restrict the weld cooling rate. Consumable developers have foreseen this and wires have been developed which 'buffer' the microstructure against compositional changes arising from dilution. These can now outperform solid wires in many single and two-pass welding applications.

Non-acicular microstructures can also give good toughness if they are properly designed though, to date, among tubular wires only self-shielded types have made serious use of such structures. Future limitations on yield-to-ultimate strength ratios for weld metals could see this situation change.

The root runs of a multi-pass weld often have the poorest toughness. When using a gas-shielded process, it is difficult to provide adequate shielding for the back of the weld and so nitrogen may be picked up from the air. Unfortunately, nitrogen is one of the elements responsible for dynamic strain ageing, which causes embrittlement of steel that is deformed while warm. The weld root is susceptible to strain ageing if later passes produce deformation in the form of rotation about the root. Unrestrained procedure test plates are especially prone to this and may show poor root toughness that would not be reproduced in a more rigid structure, so restraining test plates to prevent 'butterfly' distortion is a worthwhile precaution. Dilution of the root pass, though not high, can move the composition away from its optimum and add to any degradation of properties. Many fabricators prefer to remove the root by back-gouging to a specified radius of 5 or 6 mm to remove any embrittled material.

## Post-weld heat treatment

Post-weld heat treatment is often referred to as stress relief because its main purpose in structural steels is to relax residual stresses in the joint and thereby improve its resistance to brittle fracture. Removal of hydrogen is a further benefit. In non-alloyed and micro-alloyed weld metals, however, some metallurgical processes may take place which lead to embrittlement. Grain boundary carbides may thicken to the point where they can become cleavage crack initiators.[27] Segregation of phosphorus, arsenic and antimony to grain boundaries can cause reversible temper embrittlement.[28] When testing weld metals to see whether they will be suitable for use in the stress-relieved condition, it is important to ensure that the heat treatment time and cooling rate as well as the holding temperature are representative of those of the actual fabrication. Because of batch-to-batch variations in residual elements in welding consumables, results from a single batch must be viewed with caution. Tests on thinner sections or with faster cooling rates may not show the maximum embrittlement. In general, basic flux-cored wires are likely to be the safest choice for stress-relieved applications but good results are claimed for some rutile wires and manufacturers will advise on how to use these.

## Troubleshooting

Most welding problems can be avoided by good planning and the sections above outline some precautions which may help, but sometimes trouble seems to come out of the blue and a diagnosis must be attempted.

### Arc instability – poor wire feeding – electrical contact problems

Welders from time to time report conditions described as arc instability and sticking or snatching of the wire in the torch. These may be caused by poor wire feeding, by poor electrical contact between the wire and the torch tip or, more rarely, by other wire surface phenomena. The difficulty for the welder is to distinguish between the possible causes of the problem.

The first approach should be to check for feeding difficulties.

An extreme condition is seen when the wire judders when being fed through the torch with no arc. After verifying that the correct tip size is being used, conduits and torches should be checked for build-up of debris from the wire surface. Manufacturers usually coat the wire with some form of lubricant to ease its passage through the conduit and if this is in solid form, for example a graphite mixture, an accumulation in the conduit may occur. Swarf under the wire feeder drive rolls is an indication that the wire is meeting resistance and may be a sign that the rolls are over-tightened and causing the wire seam to open and spill powder. Drawn wires of small diameter are more resistant to excess pressure than rolled types, especially those of larger diameter. Rolls should be checked to ensure that they are matched to the wire size. Liners should be of the correct length: problems can arise if the liner is too short and can move longitudinally within the cable.

Having checked the equipment, the wire should be inspected for kinks and waves by running a sample between finger and thumb before it has been through the drive rolls. A badly butted seam feels rough as do slivers created during manufacture, and any roughness will result in damage to liners and shorter tip life. Wiping the wire with a clean tissue will reveal how clean the wire surface is, but experience is needed in assessing wires lubricated with graphite since these may appear dirtier than they are. Clogging of the liner and torch with graphite shows that an excess is present. At the other extreme, the wire may not have received enough lubricant to ensure proper feeding. Most manufacturers perform feeding tests regularly on their products, often by measuring the feeding force needed to push the wire through a conduit coiled into a standard radius. They will advise users whether a solid or liquid lubricant is used and how to evaluate the wire when other means of improving feeding have been exhausted.

When attempting to feed wire through unusually long conduits, for example those of 25 m sometimes found in shipyards, very soft wires may prove impracticable. Stiff wires are capable of good results if the free diameter (cast) of the wire as it comes off the reel is not too small. Manufacturers supplying wire on spools generally aim for a cast not very much larger than the spool diameter so that the wire does not fly off the reel like an uncoiling spring if the end is accidentally released. Wire supplied on larger coils or bobbins will have a larger free diameter and packs of 150 kg or more are now available from which the wire is straight when it emerges. These not only give improved feeding, but eliminate arc wander due to wire corkscrewing in the stickout.

Sometimes the arc appears unstable and the arc length variable when there are no other signs of poor wire feeding. In this case, poor contact between the wire and the tip is a likely cause. This can arise when wire that is quite straight is used with a gun with little curvature, especially in robotic applications. A more curved torch body forces the wire into better contact with the tip. Wires do vary in their surface resistance and some manufacturers monitor this. Copper-coated wires have the lowest resistance, followed by wires manufactured without solid lubricants, whether baked or not. Drawn wires which have been heavily soaped in manufacture tend to have a high surface resistance. However, baked soap residues can have a beneficial effect on arc stability, as mentioned in Chapter 2, especially with electrode negative polarity. If the welder sometimes sees a conical arc and a sharp wire tip but sometimes a more 'fluffy' or cylindrical arc with a blunt wire tip without changing the welding parameters, or if the arc has a tendency to climb up the wire in search of a stable root, the wire surface may be suspect and the supplier may be asked to advise.

### Excessive tip wear

Any of the above problems may be linked with excessive tip wear or it may appear without other adverse effects. A common reaction to poor wire feeding is to increase the pressure on the wire drive rolls. If these are knurled, the wire may be indented to form a file which then attacks the contact tip. If powder spills from the wire, similar abrasion may occur. A wire which is delivered with a rough surface may wear the tip without necessarily giving poor feeding. Poor electrical contact will result in the tip running hotter and possibly softening, again leading to increased wear.

The rôle of the tip itself cannot be ignored: where tips are supplied by the manufacturer of the welding equipment, he will take responsibility for their quality. However, in the absence of national or international standards for tips, if they are purchased from another supplier then dimensional tolerances and material specifications may not match those of the original equipment.

### Monitoring the process

The demands of quality standards such as ISO 9000 have always been open to different interpretations in dealing with welding, but the concept of in-process monitoring and recording is clearly here to stay and appears as an option in prEN 729 Pt2, 'Quality requirements for welding: comprehensive quality requirements'. For critical applications, on-line recording of welding parameters offers extra reassurance that written procedures are being followed. Because flux-cored wire can be used over a wider range of parameters than most other types of consumable while still producing superficially acceptable welds, it is particularly important to monitor the process if optimum metallurgical properties are to be achieved.

Monitors recording current, voltage and wire feed speed digitally on paper have been available since the mid-1970s and are in widespread use. Newer systems record either directly on to a microcomputer or use a solid-state memory in the form of a 'smart card' which can be down-loaded to a computer at the end of a shift.[29] In one version of the latter system, Fig. 4.10, the desired parameters can be entered initially from a PC. The monitor then interfaces with the power source to control the voltage and wire feed speed and at the conclusion of welding the parameters recorded may be compared with the preset values. A particular advantage of monitoring both the wire feed speed and the current is that their ratio is a good indicator of effective stickout. Indeed, some installations where stickout is critical have used this to sound a buzzer as a signal to the welder.

There is unfortunately no monitor commercially available at present to check that adequate shielding gas is present during welding, although equipment has been demonstrated using radio frequency emissions from the arc and results appeared promising.[30] More recently, spectroscopic detection of water vapour in the arc has been reported and proposed not only as an indicator of weld hydrogen levels but also as a test of shielding.[31] Until such time as equipment is marketed, regular checking of gas flow by

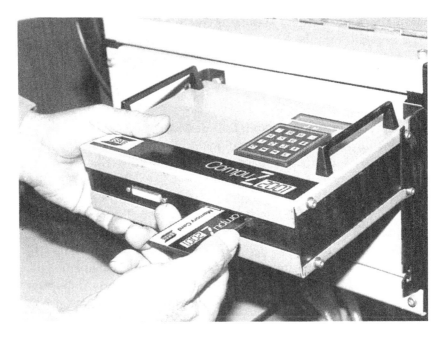

**4.10** Microprocessor monitor using a 'smart card'.

means of a gauge with a cup which fits over the torch end is recommended.

### Training and technique

Because of the ease with which welders who are used to other processes, or even non-welders, can quickly produce apparently good welds with tubular wire, some users have believed that only minimal training in the use of the process was needed. But if the quality of the welds is to be more than skin deep, proper instruction is essential. Some elements of theory, training in setting up the equipment and practice in the manipulative techniques needed for welding with different products in different positions are all important.

The theory of semi-automatic welding in general is well covered in textbooks and other documents[19,32-34] and some additional considerations relating to tubular wire are dealt with in other chapters of the present book. The detail necessary for a welder to perform efficiently will depend on the range of tasks he has to deal with and the degree of supervision. Setting up and adjusting equipment is not covered so well and even within companies making welding equipment and consumables, difficulties are sometimes experienced. Manipulation of the torch is best learned in the workshop and only general guidance will be offered here, but the training videos now available can be helpful.

A checklist for setting up to weld with tubular wire should include ensuring that the wire reel is securely attached, both for the safety of the operator and so that the over-run brake operates, that the rolls are correctly and not over-tightened, that the cable is in good condition and not kinked and that the polarity is as specified by the manufacturer or the welding procedure. Where the equipment offers the possibility to choose different shroud lengths, the length should be selected to allow an electrode extension appropriate to the application: for downhand welding at high currents, the stickout may be 25–30 mm and a long shroud will extend beyond the contact tip. For positional welding with a stickout of 10–12 mm, on the other hand, a short shroud will allow the tip to stand a few millimetres proud of the nozzle. An appropriate gas flow should be selected as described earlier and checked at the torch.

Where no guidance is available for selection of welding parameters, the type of wire will determine the starting point. Rutile wires are the easiest to set up because they show no transition in metal transfer behaviour over a wide range of currents. A 1.2 mm rutile wire should operate from about 140 to 300 A in spray transfer. The current is set by means of the wire feed speed control, which is often mounted on the wire feed unit. The voltage will be set high enough so that a fine spray is achieved – too low a voltage will result in a globular transfer – but not so high that the arc length increases beyond 3–4 mm. Higher voltages will reduce penetration and may lead to undercut at the weld edges. Different wires will need different voltages, but these are likely to be in the range 20–32 V with argon-based gases. Most wires will need 1–2 V more when welding under $CO_2$. Wires intended for downhand welding (EX0T-1) will readily run at high currents, especially with $CO_2$. Wires of 1.6 mm may achieve 450 A and at 2.4 mm, 650 A is possible. When using all-positional wires in the downhand position, the stiffer slag may lead to a deterioration in weld surface finish at the upper end of the current range and gas flats may appear unless minimum stickout recommendations are followed. Thus in practice a 1.2 mm E71T-1 wire would probably be used at 280 A whereas an E70T-1 type might be used at over 300 A. In welding vertically upwards, E71T-1 wires would typically be comfortable to use at 180 A though 250 A is possible with certain types on heavier plate and still higher currents in mechanised operation. Rutile wires are always connected with electrode positive.

To set up metal-cored or basic wires to run in spray transfer, the approach is similar. However, while a wide range of currents may be used with rutile wires, metal-cored wires will fall into a globular transfer mode at currents below the optimum. Some users have attempted to insist on a wide operating range so that they can run a 1.2 mm wire at, for example, 250 A and some manufacturers have made wires to do this, but there are always compromises to be made in achieving this and such wires are rarely competitive at the higher currents where metal-cored wires come into their own. A more practical approach is to run the wire where possible at a high current, for example 300–350 A on a 1.2 mm wire, and to adjust the weld size by varying the welding speed. This will produce high quality welds with little spatter.

It may be noted that the more skilled the welder, the lower the voltage he will tend to select for a metal-cored wire in spray, so that an occasional crackle is heard as the wire short-circuits: this may be to provide a degree of aural feedback in response to movements of the torch. Basic wires do not produce a pure spray transfer and there will always be a slight crackle in the arc. The most difficulty in setting up, however, is found when basic and metal-cored wires are to be run in dip transfer. The tolerance boxes for this are usually quite narrow and there is no continuum between the dip and spray areas.

In principle, the first stage in setting up to weld in the short-circuiting mode is to set a wire feed speed and hence current within the required range – perhaps 80 to 180 A for a 1.2 mm wire. The voltage is chosen so that the wire does not 'stub-in' to the pool because the voltage is too low, nor is the arc length too great. A higher voltage will increase the risk of undercut but correspondingly improve sidewall fusion. Metal-cored wires operate at lower voltages than solid wires, typically 15–20 V, and welders unfamiliar with tubular wires may find them difficult to tune in at first. It is often helpful to start setting up with a low inductance, increasing this if necessary to reduce spatter when the voltage has been optimised. The final value of inductance chosen will usually be lower than for solid wires: higher values lead to a condition welders describe as too hot. In welding with solid wires, it has generally been considered that the higher the short-circuiting frequency the better, and most manufacturers have followed this precept in designing tubular wires. However, certain basic wires are apparently designed to mimic the metal transfer behaviour of popular low-hydrogen MMA electrodes and must be set up to give a rather low dip frequency. There is no substitute for the guidance of a skilled instructor if welders are to become proficient in the use of basic and metal-cored wires. The use of electrode negative polarity will produce a finer droplet size and is preferred for positional welding provided arc stability is acceptable: as mentioned earlier, the surface of baked wires is more conducive to a stable cathode than that of shiny wires. At the higher currents used for spray transfer, many basic and metal-cored wires perform equally well on either polarity. Manufacturers are obliged by national and international standards to provide information on optimum current, voltage and polarity for their products but their recommendations are necessarily conservative.

Makers of self-shielded wires, perhaps more conscious than others of the particular techniques and skills involved in using these, have generally not only provided precise information on operating parameters but have often run welding schools to train welders and supervisors. A key factor in translating the manufacturer's data into sound welds is the use of equipment which displays the wire feed speed. This, together with accurate control of voltages which are generally lower than users of gas-shielded wires will be accustomed to, provides the basis for success. Maintenance of the correct stickout during welding is an equally important factor, and this will be verified if the arc draws the expected current for the wire feed speed set.

The major part of training in the use of tubular wires will be concerned with the welding operation itself and torch manipulation. Rutile wires are again the easiest starting point. In horizontal-vertical (1F) fillet welding, the torch should be held at an angle of 45° to the horizontal and vertical plates and a backhand or 'pulling' technique is employed: that is, the wire points back towards the completed bead, the torch making an angle of 60–70° to the weld line. In this way the arc force prevents the slag from running in front of the weld and the danger of slag traps is reduced. Only on small fillets made at high speed, where the speed of the leading edge of the pool keeps it ahead of the slag, is a forehand or 'pushing' technique recommended with wires containing flux. In that case, the result is a less convex bead shape. The tip of the wire in fillet welding points at the bottom plate at a distance between about 1 and 1.5 diameters from the joint line, Fig. 4.11.

Manufacturers provide guidance on the appropriate stickout to use and in downhand fillet welding with gas-shielded wires this is likely to be in the range 20–30 mm. Shortening the stickout increases the penetration and reduces the convexity of the bead. Too short a stickout may lead to a rapid build-up of spatter in the nozzle but if the stickout is too long there may be a loss of gas shield.

Basic wires are handled in much the same way for HV fillets but metal-cored wires are generally used with a forehand technique for all but the largest fillets since the slag volume is small. The torch is held rather more vertically than in backhand welding, making an angle of 70–80° with the weld line. This produces an attractive mitred fillet. The fillet size can be controlled by varying the welding speed. For large fillets or multi-pass butt welds, a backhand technique gives better penetration and gas coverage.

If a fillet weld of more than about 10 mm leg length (7 mm throat thickness) is required, more than one pass will be needed. In that case, the first pass is made as before and the next is made with the wire tip aimed at the toe of the first bead with the torch held at a rather more acute angle to the vertical plate. The third run is at the upper edge of the first, this time with the torch held a little closer to the horizontal. If more than three runs are needed, each layer is built up from the bottom in the same way. Although there is a modern tendency to make large fillets with small diameter wires, this leads to an increased risk of lack-of-fusion defects and the practice should not be encouraged.

Where it is possible to position the joint so that the weld can be made in the flat or gravity position, larger fillets can be made with fewer runs by using a weaving technique. The torch is angled forward at 60–70° to the joint line, with the wire pointing back

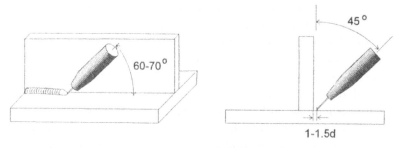

**4.11** Welding technique for HV fillets using rutile wires.

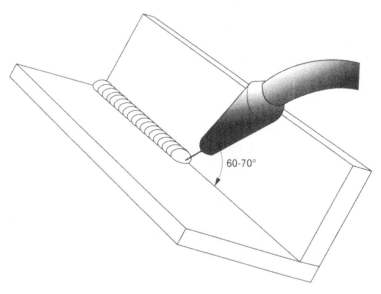

**4.12** Welding technique for flat (gravity) fillets.

towards the completed weld and aimed at the front edge of the pool, Fig. 4.12. A weave width of up to 20–25 mm may be used and the finished weld should have a flat, smooth surface with good penetration beyond the root. Not all wires are equally suited to making large gravity fillets: some basic wires in particular have a fluid slag which tends to run ahead of the pool despite the arc force pushing it backwards.

For butt welds made in the flat position, only metal-cored wires run under argon-rich shielding gas or some self-shielded types would normally be used to weld into an open root. With a feather-edge preparation and a gap of 1.5–3 mm, or with a 1.5 mm root face and 3 mm gap, wire of 1.2 mm diameter used in the short-circuiting regime at 100–140 A and 15–18 V can produce a good profile at the back of the joint though training and experience are needed to ensure adequate penetration with no lack-of-fusion defects. Otherwise, either a permanent or a removable backing must be used as described in Chapter 3. Welding techniques are similar to those for fillet welding but in most cases, weaving may be introduced as the joint progresses. If it were not for the need to restrict heat input to ensure adequate mechanical properties, each layer could be made using a single run by weaving across the complete width of the joint. Where very severe heat input restrictions apply, 'stringer' beads with little or no weaving must be used. In practice, requirements will generally be met when the first one or two runs are of full width and the second or third and later layers are made with two or more runs each using a weave of half the joint width. This is known as a 'split weave' technique, Fig. 4.13. Procedure tests and manufacturers' recommendations on heat input will establish how early in the joint the weave needs to be split and whether more than two runs per layer are necessary.

When welding vertically with flux-cored wire, 1.2 mm is the most popular size although 1.4 mm is sometimes used where the application calls for a high proportion of downhand welding with a limited need for positional work. Many self-shielded wires are used positionally in larger diameters, often between 1.7 and 2.0 mm, since they are difficult to make in smaller sizes and their formulations have been optimised to give satisfactory arc charac-teristics at low current densities and voltages.

Rutile wires are used vertically in spray transfer at currents which may range up to 250 A on heavy plate though 180 A is more common. For all-positional welding with gas-shielded wires a rather short stickout of around 12 mm is recommended and the use of a short shroud helps access and visibility. A single pass fillet may be made vertically up without weaving, the torch being held pointing upwards at about 10° to the horizontal and the travel speed adjusted to direct the arc towards the leading edge of the pool. Where a leg length of more than 6 mm (throat thickness 4 mm) is required, it is normal to use a multi-pass technique. The

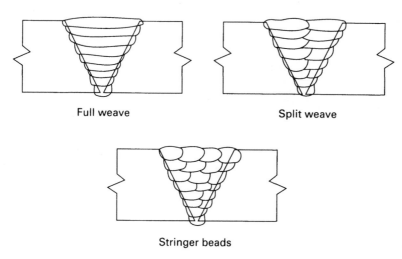

Full weave                    Split weave

Stringer beads

**4.13** Bead sequences for multi-pass welds.

second and later passes are made with a side-to-side weave and with a good EX1T-1 wire, a flat weld profile should result.

Overhead fillets are made with somewhat higher currents, for the same plate thickness, than vertical-up welds. For semi-automatic welding a backhand technique is preferred and the fillet should again have a flat profile.

Special provision must be made for the root of vertical butt welds made with rutile wires and ceramic backing tiles are one of the best methods of achieving reliable results. The root is made with a 'half-moon' weave pattern to ensure the joint edges are properly fused. Subsequent passes are made with a side-to-side weave. Unless mechanical property requirements limit the heat input, a single weave can be used up to a width of 20–25 mm, followed by a split weave.

Positional welds with basic and metal-cored wires are made in the dip transfer mode. The parameters must be set precisely, as described above. For fillet welds the torch is again held 10° below the horizontal but in this case a slight weave in a triangular pattern is usually necessary. Later passes use a side-to-side weave. Because the current is lower than with rutile wires, it is helpful to pause rather longer at the edges of the weave to ensure proper fusion. Vertical welds made with basic wires in particular tend to be more convex than those made with all-positional rutile wires.

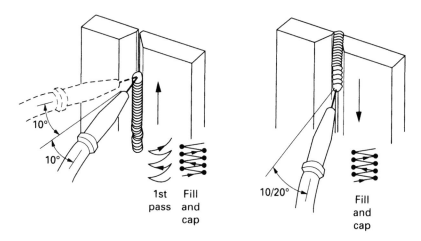

**4.14** Welding techniques for vertical butt welds with basic and metal-cored wires.

It is possible with care to weld open roots with basic and metal-cored wires. The torch should now point downwards at 10° to the horizontal and the arc is directed on to the pool itself to limit burn-through, Fig. 4.14. A half-moon weave pattern is used. Following passes are made as for fillet welds. Again, the point at which a split weave needs to be used will be determined by mechanical property requirements.

It is difficult to generalise about techniques for positional welding with self-shielded wires. Types which do not contain barium may need an arc voltage up to 25 V to produce flat single-run fillet welds and will do so with the torch held horizontally. Barium-containing wires operate with much lower voltages and usually with the torch held below the horizontal. It is most important to follow the manufacturer's recommendations on stickout, which may be a little longer than for gas-shielded wires.

# CHAPTER 5

# Process economics

The economic case for using tubular wire is often so clear that no sophisticated costing methods are needed to prove it. The comparison with the MMA process, for instance, rarely turns on figures after the decimal point and it is intuitively clear that semi-automatic and mechanised processes offer great potential for productivity increases. Where these processes are already being used with solid wire, detailed costing may be needed to show the benefit of changing to tubular wire and some comments on this follow. Some productivity gains are not reflected in the cost per metre of weld as conventionally calculated but may still be the best economic argument for a process change. Quality improvements certainly have an economic value but are often only appreciated after the process change has been made.

This chapter briefly presents a number of examples of the successful application of tubular wire, and shows in each case how the use of or change to the process was justified. Many examples have appeared in the literature[35-44] and the reader is referred to these for fuller details. Some examples have been included where the case was originally made without quantitative data, while others are given with costings and in currencies that were appropriate to the user: thus it has not been possible to present the examples in a standard format. It is hoped that readers will find some that are relevant to their own interests.

*Example 1: welded crane beam*

**5.1** Welding crane beam.

Beams for 130 tonne capacity overhead cranes were fabricated using large diameter, high recovery (E7024) MMA electrodes. Preheating was applied manually and the job took 5 hours per beam. Welding was mechanised using two machines to weld both sides of the beam at once and preheating torches were mounted on the same carriage ahead of the arc. Then a 1.6 mm metal-cored wire was used. The welding time was reduced to 18 min and the weld quality improved, with better penetration and no re-starts.

*Example 2: blast furnace half-jacket*

**5.2** Welding blast furnace.

Blast furnace jackets were made in medium-carbon steel up to 150 mm thick using E7018 electrodes which were re-dried before use. Extra supervision was required to maintain preheat temperatures, electrode drying and quality in general. A change was made to a basic flux-cored wire which gave a hydrogen level of <5 ml/100 g without redrying. The weld heat input increased and the preheat temperature was lowered by 100 °C. The deposition rate also increased, supervision levels could be reduced and the overall saving was estimated at 36%.

Another blast furnace was successfully rebuilt using both gas-shielded, rutile all-positional wire and self-shielded wire. A total of 15 te of consumables were used and a zero defect rate was reported.[38,41]

The above examples comparing MMA with tubular wire welding are typical in showing a clear benefit in changing process even before detailed costing is carried out. The next example uses a simple cost estimate to show the same benefit more dramatically.

*Example 3: offshore platforms*

**5.3** Forties field platform.

At the outset of platform construction for the North Sea, MMA electrodes and flux-cored wires were used in parallel. This provided an opportunity to assess their relative economic performance under realistic conditions and the following costing was made for typical vertical butt welding parameters.[43] A small arithmetical correction has been made to the published calculation.

*Table 5.1*

---

Weld costing

| Consumable | MMA | Self-shielded FCW |
|---|---|---|
| Diameter, mm | 3.2 & 4 | 2 |
| Melt-off rate, kg/h | 1.8 | 2.3 |
| Consumable efficiency, % | 60 | 80 |
| Duty cycle, % | 25 | 40 |
| Consumable price, $/kg | 1.32 | 2.70 |
| Labour and overhead, $/h | 15 | 15 |
| Deposition rate, kg/h | 0.27 | 0.74 |
| Consumable cost, $ per kg weld metal | 2.20 | 3.38 |
| Labour cost, $ per kg weld metal | 55.56 | 20.38 |
| Total cost, $ per kg weld metal | 57.76 | 23.76 |

*Example 4: automotive wishbones*

An automotive component manufacturer makes wishbone assemblies by welding together two 3 mm thick pressings. The welding is done by robots and the total weld length is 1.25 m per unit.

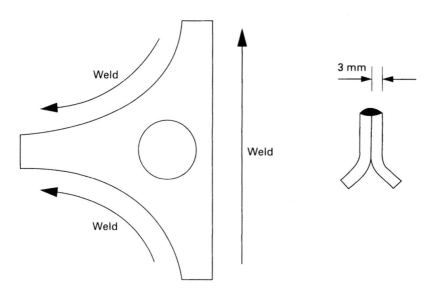

**5.4** Fabricated wishbone assembly.

Using solid wire at 180 A, 22 V the welding speed was 0.72 m/min and the spatter level was high. Changing to a 1.0 mm metal-cored wire and inclining the component so that the weld was always made downhill at an angle of 30°, the current was raised to 310 A at 28–29 V. The shielding gas remained 80% Ar, 20% $CO_2$, the stickout was 15 mm and the torch was perpendicular to the joint. Under these conditions, the welding speed could be increased to 1.32 m/min, while spatter decreased and joint consistency and tolerance to fit-up improved. Attempts were made to reproduce these results using the cheaper solid wire, but to date consistent results have not been achieved.

*Example 5: water heater cylinders*

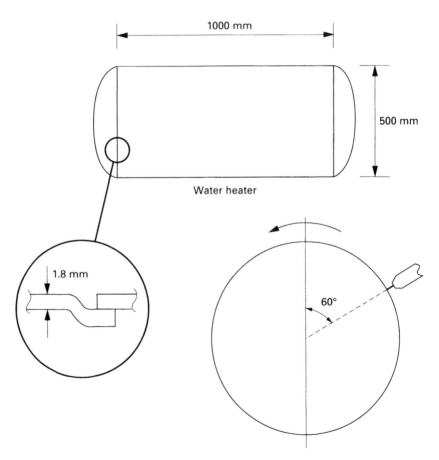

**5.5** Welding water heater.

Water heater cylinders with a wall thickness of 1.8 mm and a diameter of 500 mm, Fig. 5.5, were welded with solid wire at 400 A. Using metal-cored wire at 450 A, 32 V and 25 mm stickout on a mechanised setup, the welding speed was 3.45 m/min to give a welding time of 27 s and an overall cycle time of 48 s, some 30% faster than with solid wire. Two heads were use to weld both dished ends simultaneously. The wire feed speed was 11.8 m/min and the deposition rate was 10.2 kg/h.

*Example 6: water tube boilers*

**5.6** Welding of boiler.

Boilers, Fig. 5.6, were being welded with conventional flux-cored wire for all joints except the shell strake, for which submerged arc was used. Changing to gas-shielded metal-cored wire removed the need for deslagging in the multi-pass welds and reduced welding time by 15%.

## Costing a joint

Compared with the uncertainties of attempting to predict the effects of hydrogen in weld metal, the questions of costing welded joints, comparing processes and evaluating investment for process change might seem based on altogether firmer ground. In fact, a comparison of the approaches used by a random selection of fabricators suggests that this view may be optimistic. At a Welding Institute conference in 1971, some approaches to weld costing were described[45,46] but difficulties were also pointed out.[47]

There is now available a variety of software for costing welds and as well as bringing some uniformity of treatment to the subject it serves to make explicit some of the assumptions which often pass unnoticed in cost equations. However, even commercial software can include doubtful propositions: a version widely used in the 1980s assumed that the shielding gas flowed continuously, thus inflating the final gas cost, for example by a factor of 5 for a 20% duty cycle. A check with a pocket calculator is quickly done and a typical costing algorithm is shown in Fig. 5.7.

The first step is to calculate the joint volume and to convert this into a weight w of deposited metal, say in kg/m of joint length. A nominal electrode efficiency $\eta$ will be applied to work out the weight of consumables needed. For tubular wires, this could be 95% or more in the case of metal-cored wires, 85–90% for gas-shielded flux-cored wires and 75–80% for self-shielded types. MMA electrodes can reach efficiencies of just over 80% for thinly-coated cellulosic types as measured by the ratio of filler metal deposited to electrode consumed, but the stub ends which are not consumed still have to be paid for and a ratio of 50% of deposited metal to purchased electrodes would be typical of a well-run site. From these parameters the weight of consumable required per metre of weld is calculated as $w/\eta$.

To the cost of the welding consumables themselves must be added those for electricity and shielding gas. An electricity consumption of 1¾ kWh/lb (3.85 kWh/kg) of deposited metal for all fusion welding processes used to be quoted, but the graphs in Chapters 1 and 3 suggest that for flux-cored wire welding values of 1.5–2 kWh/kg are more appropriate. This probably reflects the relative inefficiency of older power sources, which often consumed

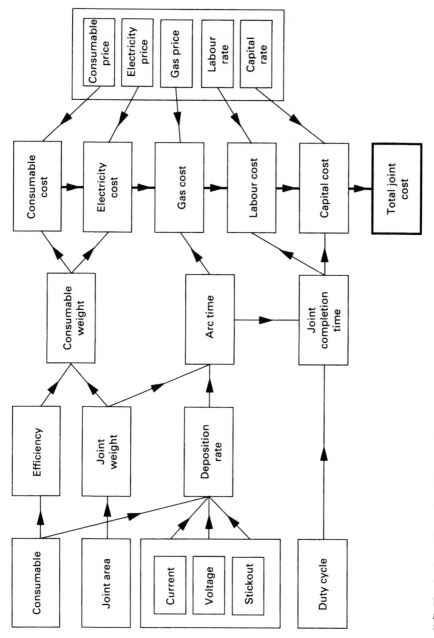

**5.7** Typical costing algorithm calculation.

2 kVA or more for an output of 1 kVA. Modern power sources can have efficiencies of 80–90%. Recommended gas flow rates range from 12 l/min at 100 A to 20 l/min at 400 A for argon-based mixtures. Flow rates for $CO_2$ may be 25% less, as the denser gas is more resistant to disturbances. The ratio of shielding gas weight to deposited metal weight varies from about 50% at low currents to 15–20% at high currents, but given the use of purging, inching the wire and the need to return gas bottles before they are completely empty, many fabricators find gas usage approaches wire usage weight for weight.

From the weight of metal in the joint, arc times can be worked out if deposition rates are known. Manufacturers publish this information in forms which can be readily used for costing calculations and stopwatch measurements in the field confirm the accuracy of predictions provided the welding parameters remain as specified. Unauthorised knob-twiddling can increase or decrease times if training and supervision are inadequate – higher currents may buy productivity at the expense of weld properties.

The most difficult aspect of costing, and the one which offers the greatest scope for wishful thinking, is the estimation of operator factor or duty cycle, F. Some costing software offers 25% for MMA welding and 40% for semi-automatic welding as default values but actual figures vary widely from industry to industry and shop to shop. Welding manufacturers sometimes use a figure of 1 te of wire per year as a typical consumption for a semi-automatic welder: at an average deposition rate of 2 kg/h, this would imply an operator factor of about 25%. On the other hand particular sectors, such as the pipeline and earth-moving equipment industries, have a tradition of achieving much higher factors. Fabricators can very readily estimate existing duty cycles from consumable consumption and increasingly from records of dataloggers used for monitoring welding parameters, assuming industrial relations permit this. Predicting the result of a process change is more difficult. In the 1970s, shipyards bought semi-automatic equipment despite agreements that there would be no reduction in the numbers of welders, yet higher duty cycles formed part of the case for the purchase. When flux-cored wire was used to rebuild a blast furnace, operator factors were initially no higher than those of the previous MMA process until MMA sets were introduced as a stand-by for use if the semi-automatic sets broke down; at that point, breakdowns fell and duty cycles rose. The higher the inherent productivity of the process and the greater the capital cost involved, the more important duty cycle becomes. With the right logistic support, semi-automatic welding is capable of very high duty cycles. Management must ask at this stage where cost benefits will come from organisationally: how many welders will be redeployed or whether increased output will take up the slack. Failure to address these questions has all too often led to disappointment if not disillusion with a change to flux-cored wire welding.

With arc time and duty cycle known, the labour cost $C_L$ can be calculated using the appropriate hourly rate. This is normally regarded as straightforward given that welding engineers are not generally in a position to influence accounting conventions used, but as the attribution of overheads is one of the more interesting and controversial questions in the theology of cost accounting, they should at least be aware that the convention chosen can greatly alter the predicted effect of process change. In particular, when a large fixed element is added as overhead to the labour cost of a small team, there is a danger of overestimating the marginal cost or benefit of manning changes: the overhead may in practice have to be carried at the same level but redistributed over more or fewer workers.

From the hourly labour cost L the labour cost per metre is calculated as $wL/\eta F$. This will nearly always be the largest item in the costing and is the source of the greatest savings when changing from another process to tubular wire welding.

The last item in the conventional cost equation is capital depreciation. This is typically arrived at by depreciating the equipment over a fixed period and applying a suitable operating factor to obtain the cost per hour. Thus £1000 written off over 5 years of 1800 h for equipment with a duty cycle of 30% costs £0.37/h in depreciation. The inclusion of interest costs, for example by discount costing, and in inflationary times the use of replacement cost accounting increase the figures somewhat but it is often claimed by the advocates of semi-automatic welding that the extra cost of the equipment is offset even within the depreciation calculation by its higher operating factor.

When the joint cost is arrived at by summing the cost of consumables, gas, electricity, labour with overheads and depreciation, the result may be quite suitable for estimation purposes and generally forms the basis for decisions on process change. It must be remembered, however, that although users can measure consumable efficiencies or deposition rates to within a percentage point or two, the multiplying factors of duty cycle and overhead rate are much more difficult to quantify, especially when a change of process is in view.

*Example 7: flat butt joint*

This Japanese example[44], Table 5.2, uses a 2.0 mm metal-cored wire to compare with a 1.6 mm solid wire. The practice of increasing the diameter to reflect the lower wire density when moving from a solid to a tubular wire is realistic for downhand welding but is now less common in Europe. The basis on which the duty cycle increased on changing consumable was not made clear but the overall result would be similar if it had remained the same.

*Example 8: digger beams*

Digger side beams were made with partially penetrating asymmetric V welds, the second pass in the form of a fillet. Solid wire was used to make welds of different sizes and the effect of substituting either a metal-cored or a rutile flux-cored (E70T-1) wire is shown in Table 5.3 for the case of a bead of 32.5 mm² cross-section.

Here the 31% increase in overall productivity was the main argument for changing to a metal-cored wire, although there was a small cost saving as well. The current chosen for the 2.4 mm rutile wire could probably have been increased to make this option more attractive.

## Measures of productivity

Where the available resources are flexible and demand for products high, arguments based on cost per metre of weld as described above may be sufficient to assess the viability of a proposed process change. In some other cases, productivity must be considered as a separate question.

If a pipeline contractor lays one pipeline at a time, starting from one end and continuing to the other, the speed with which a

*Table 5.2*

**Weld costing**

| | Option 1 | Option 2 |
|---|---|---|
| Joint area, mm$^2$ | 53 | 53 |
| Consumable | Solid wire E70S-6 1.6 mm | Metal-cored wire 2.0 mm |
| Current, A | 400 | 500 |
| Voltage, V | 35 | 42 |
| Shielding gas | $CO_2$ | $CO_2$ |
| Duty cycle, % | 50 | 55 |
| Consumable price, $/kg | 1.30 | 1.74 |
| Gas price, $/m$^3$ | 1.15 | 1.15 |
| Electricity price, $/kWh | 0.08 | 0 |
| Labour rate, $/h | 13 | 13 |
| Capital rate, $/h | 0.46 | 0.46 |
| Deposition rate, kg/h | 7.7 | 11.4 |
| Deposition efficiency, % | 95 | 95 |
| Joint weight, kg/m | 0.413 | 0.413 |
| Consumable weight, kg/m | 0.435 | 0.435 |
| Arc time, min | 3.23 | 2.18 |
| Joint completion time, min | 6.46 | 3.96 |
| Consumable cost, $/m | 0.57 | 0.76 |
| Gas cost, $/m | 0.09 | 0.06 |
| Electricity cost, $/m | 0.06 | 0.06 |
| Labour cost, $/m | 1.40 | 0.86 |
| Capital cost, $/m | 0.05 | 0.03 |
| Joint cost, $/m | 2.17 | 1.77 |
| Productivity, m/h | 9.29 | 15.17 |

*Table 5.3*

| Weld costing – digger beams | | | | |
|---|---|---|---|---|
| | Option 1 | Option 2 | Option 3 | Option 4 |
| Joint area, mm$^2$ | 32.5 | 32.5 | 32.5 | 32.5 |
| Consumable | A18 1.6 mm solid wire | Metal-cored 1.6 mm | Rutile FCW, 1.6 mm | Rutile FCW, 2.4 mm |
| Current, A | 400 | 420 | 400 | 450 |
| Voltage, V | 33 | 35v(−) | 38 | 34 |
| Stickout, mm | 20 | 20 | 20 | 30 |
| Shielding gas | 80/20 | 80/20 | $CO_2$ | $CO_2$ |
| Duty cycle, % | 40 | 40 | 40 | 40 |
| Consumable price, £/kg | 0.57 | 1.89 | 2.01 | 1.86 |
| Gas price, £/m$^3$ | 0.80 | 0.80 | 0.40 | 0.40 |
| Electricity price £/kWh | 0.05 | 0.05 | 0.05 | 0.05 |
| Labour rate, £/h | 25 | 25 | 25 | 25 |
| Capital rate, £/h | | | | |
| Deposition rate, kg/h | 6.7 | 8.8 | 7.8 | 7.3 |
| Deposition efficiency, % | 95 | 95 | 85 | 85 |
| Joint weight, kg/m | 0.254 | 0.254 | 0.254 | 0.254 |
| Consumable weight, kg/m | 0.267 | 0.267 | 0.298 | 0.298 |
| Arc time, min | 2.27 | 1.73 | 1.95 | 2.08 |
| Joint completion time, min | 5.68 | 4.32 | 4.88 | 5.21 |
| Consumable cost, £/m | 0.15 | 0.50 | 0.60 | 0.55 |
| Gas cost, £/m | 0.04 | 0.03 | 0.02 | 0.02 |
| Electricity cost, £/kWh | 0.02 | 0.02 | 0.03 | 0.03 |
| Labour cost, £/m | 2.36 | 1.80 | 2.03 | 2.17 |
| Capital cost, m/h | 0.00 | 0.00 | 0.00 | 0.00 |
| Joint cost, £/m | 2.58 | 2.36 | 2.67 | 2.77 |
| Productivity, m/h | 10.57 | 13.89 | 12.31 | 11.52 |

joint can be made and lowered off determines the total time needed to complete the work. The 1–4 welders who make the root pass could be considered the only direct labour involved, and the whole cost of running the company falls on them as overhead. The linear speed at which the root run can be deposited is the key to the number of butts that can be completed in a shift and so to the profitability of the project. Tubular wires can score over cellulosic electrodes if they allow lowering off after the root pass without having to wait for the hot pass to be completed. The message here is that when welding is on the critical path for a manufacturing process, changes in productivity have a much greater effect on the operation than would be seen from conventional costing calculations with fixed overheads.

Another factor which conventional costing does not always adequately account for is the relationship between resources available and demand for the finished product. If a manufacturer with a fully loaded shop is offered a cheaper consumable or one which increases output, the effect on the cost per metre of weld might on a simple calculation appear the same. However, one would require further capital investment to increase output while the other allows an immediate increase. This situation often arises in considering the alternatives of solid and flux-cored wire, where the extra consumable cost of tubular wire is significant but so, too, is the potential improvement in productivity.

Increasingly, costing programs allow users to look at output in terms of welds completed or units manufactured per hour or per shift. This gives one more way of evaluating a proposed process change before funds are committed.

# CHAPTER 6

# Equipment requirements

The equipment used for welding with tubular wire is fundamentally the same as that for solid wire MIG/MAG welding, Fig. 6.1, although the self-shielded process has rather different and simpler requirements.

### Power sources

The use of constant potential power sources for MIG welding was first proposed in 1949[48] and these have generally been used for both solid and flux-cored wire ever since. Arc stability arises from the fact that the impedance of the power source is lower than the effective impedance of the arc in normal use. If the arc shortens, coupling between the power source and the arc improves, energy transfer increases and the wire melts faster until equilibrium is restored. The stringent definition of constant potential in the original patent has been slightly relaxed and a fall of up to 7 V per 100 A is now normal. Some early sets featured a variable slope for the static V-I characteristic but this is not widely available today, although the advent of microprocessor-controlled equipment is now making it possible once more.

For many years, power sources consisted of transformers with rectifiers, arc voltage being controlled by a variable primary tap. In welding with dip transfer, the rate of rise of current during the short circuit must be controlled to reduce spatter and for this purpose a variable inductance was provided. Partly for economic reasons and partly because some welders found it difficult to optimise the inductance setting, many manufacturers moved away

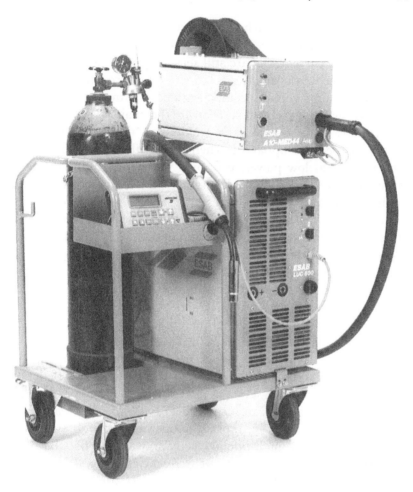

**6.1** Power source and wire feed unit for gas-shielded welding
with tubular wire.

from continuously variable inductances to two or three fixed settings, though more recently some fully variable controls have
appeared again. For flux-cored wire, welders will often select
lower inductance settings than for solid wire but this is a matter of
personal preference and the fixed settings offered will provide an
adequate choice.

A disadvantage of conventional transformer-rectifier sets was
their sensitivity to variations in mains voltage, which can be
significant in some parts of the world. In the early 1970s, welding

sets were introduced which used thyristors to rectify and regulate the current. Because a feedback signal from the arc voltage is used to control the current, the arc still sees a constant potential source and if the feedback signal is compared to a constant reference potential, mains voltage compensation is provided. Thyristor-controlled welding sets took over in the majority of industrial applications of semi-automatic welding in the 1980s. One disadvantage became apparent when sets with high ratings were used to weld at low power: the spiky waveform produced by the thyristors gave poor welding characteristics compared with older welding sets. This was not necessarily an inevitable defect of the technology, but it happened that at that time there was intense pressure on manufacturers to reduce costs. Half-bridges were cheaper than full bridges and copper for chokes was expensive, so some corners were perhaps cut. Today's thyristor sets are capable of good performance, though operation in the dip transfer mode remains the most severe test of larger sets.

Invertor technology appeared in welding sets at the end of the 1970s. The objective was to reduce the weight of the transformer by operating it at a much higher frequency than the mains. To this end, the mains supply was first rectified and then fed to a high frequency inverter. Voltage reduction could then be achieved by a light, air-cored transformer before the output was again rectified for use. Initially, most sets operated at high audio frequencies and their high-pitched scream was not popular with welders, but modern sets work at frequencies up to 100 kHz and the loudest noise is the cooling fan. An incidental benefit of the high operating frequency is the rapid response of the output to control signals which it makes possible. Together with the use of microprocessors in control circuits, this now allows great flexibility in modifying both the static and dynamic characteristics of power sources. By using the rate of change of current as a further input variable in the feedback loop, the effect of inductance can be simulated electronically, with the additional refinement that rates of rise and rates of fall of current can now be controlled independently. The latest equipment is claimed to have a response time measured in microseconds.[49]

If inverter power sources did no more than this, they would still be able to reproduce the characteristics of the best conventional welding sets and this is a common reaction of welders trying a modern unit for the first time: it welds 'like welding sets used to'. But to do only that would miss the opportunity offered by the rapid, closed loop microprocessor control. The ability to preset the way in which one parameter varies in response to another, for instance voltage in response to wire feed speed, means that 'single knob' control becomes a reality even in normal continuous current welding. The provision of pulsed current, again of course with synergic control, is only a matter of software. Even dip transfer can be improved, not just by simulating the effect of a real choke, but by sensing or predicting the start and finish of short circuits and adjusting the electrical parameters to maximise heat input and minimise spatter. Individual manufacturers all have their own control algorithms for this and to the welder, differences in software may have more effect than differences in hardware.

A further possibility offered by a welding set configured by software is the ability to work equally well with a range of processes, not only as a semi-automatic machine with solid or flux-cored wire, but also with MMA or TIG welding. The example quoted in Chapter 5 of a fabricator providing MMA sets as a backup to semi-automatic sets at every station for a job with a critical deadline could have been achieved without physical duplication if modern inverter sets had been used.

Microprocessor-controlled inverter power sources indeed provide the welding engineer with a tool capable of getting the best out of most processes and consumables. At present they are significantly more expensive than their conventional counterparts, but this margin is likely to decrease as the cost of electronic components continues to fall. As can be seen in the previous chapter, the capital cost of welding equipment does not figure very highly in joint cost calculations, certainly when set against any benefits in repair or downtime rates resulting from more precise control of welding parameters and arc characteristics.

If equipment is being bought specifically for pulsed arc welding, it must be remembered that the requirements for tubular wire are slightly different from those for solid wire. The core of a flux-cored wire may contain granular particles perhaps up to a third of the wire diameter in size, and as the wire is drawn down in manufacture, the tube wall thins unevenly on a sub-millimetre scale. Pulsed arc welding depends on melting precisely controlled amounts of wire at each pulse to form a single droplet: in the case of a 1.2 mm wire, about 1.1 mm in length would typically be melted. Variations in wall thickness mean that melting rates are uneven over a very short time scale and a system which cannot respond rapidly is not able to maintain a sufficiently constant arc length. The use of a constant potential pulse can provide this response from a passive system, or control of pulse length or separation by sensing arc voltage can do the same thing actively. Some sets of older design, however, do not offer sufficiently fast response and these cannot be recommended for use with flux-cored wires.

As tubular wires vary more widely in electrical and physical characteristics than solid wires, pre-programmed synergic para-meters optimised for one type may not work so well for another. If the welding set is to be used with consumables from its own manufacturer, the wires for which preset parameters are offered are likely to be named. In other cases, the manufacturer's advice should be sought and some fine tuning may be needed.

When synergic control of pulsed welding was being developed at The Welding Institute, it was axiomatic in their approach that the wire burn-off rate was proportional to the mean current. In fact, because melting is affected by the $I^2R$ heating of wire in the stickout, pulsing the power increases the burn-off rate by increasing the ratio of RMS (root mean square) to mean current as described in Chapter 3. This gives the designer of power sources an opportunity to alter the melting rate, or in other words the relative amounts of energy delivered to the wire and to the parent material, by changing the pulse shape. A tall, narrow pulse will produce a high RMS current compared with a low, wide pulse giving the same mean current. Changing from a square to a trapezoidal pulse of the same area has the same effect. In this way, parameters can be optimised for downhand welding, where a tall or square pulse is used for maximum deposition rate, or for vertical welding, where a flatter or ramped pulse gives more penetration into the joint. This feature is beginning to become available on microprocessor-controlled welding sets.

Despite arguments over the years about whether rated duty cycles should be measured over 5 or 10 minutes and over the minimum voltage that power sources should have to develop at rated currents, it is unlikely that problems will arise with any set supplied to a recognised national or international standard.

Purchasers hoping for a high operating factor for flux-cored wire will specify duty cycles accordingly and experience suggests that power sources survive in good order long after manufacturers could reasonably have hoped for replacement orders.

Increasing requirements for quality control during the 1980s led to the introduction of welding sets with a facility for presetting and locking the welding parameters. These could then be fixed by a supervisor, while the operator was given a trim control with which he could make fine adjustments. Usually, a number of different conditions could be preset, corresponding, for example, to downhand and vertical welding. On microprocessor-controlled equipment the number of possible settings may be large and protection by password may be provided.

Self-shielded wires, notwithstanding warnings from some of their manufacturers, run quite satisfactorily on many standard types of semi-automatic equipment. However those which contain barium may need operating voltages lower than is normal for gas-shielded wires, in some cases down to 13 V. Not all power sources can provide this and the range available should be checked. At the simplest level, self-shielded wires are sometimes run from a power source with a drooping characteristic intended for MMA welding, with the wire feeder connected in parallel with the arc. As the arc length varies, the voltage across both it and the wire feed motor varies too and the motor speeds up or slows down to restore the required value. At the low burn-off rates involved, this system works satisfactorily and provides an inexpensive option for simple fabrication and for repair and maintenance work.

In the 1960s and 1970s, self-shielded wires in Japan, Canada and some other parts of the world were often run from AC power sources. Over the years, this practice has declined and those wires designed for optimum mechanical properties are recommended for DC only, but wires are still available even in the UK which run well enough on AC, using an MMA power source and an add-on feeder as described above.

## Wire feeders

Tubular wire can be more difficult to feed than solid wire so wire feeders need to be chosen with care. Early flux-cored wires were often rolled from relatively thin strip to a final diameter of 2.4 mm or more and underwent little work hardening compared with modern small diameter drawn wires. The early wires were soft, buckled easily in the conduit and were soon flattened by over-tightened feed rolls. The design principles which allowed those wires to be fed successfully should thus ensure completely trouble-free performance with today's improved products.

Solid wires are usually fed by smooth drive rolls, one of which is grooved to provide the wedging action which generates the necessary frictional force for feeding. Feeding of tubular wires of more than 1.2 mm diameter is helped by the use of knurled rolls which can operate with lower pressures. On sets where the operator can alter the roll pressure, care should be taken not to set this too high. This increases the risk of damaging the wire and if knurled rolls indent the wire, it can act as a file and in turn damage the contact tip. If the wire fails to feed with normal pressure, something else is wrong in the system.

Many wire feeders made with flux-cored wire in mind have two pairs of drive rolls, Fig. 6.2. The benefits of this depend on the design and construction of the system. Where each set of rolls is independently driven, either by its own motor or through differential gearing, the feeding force is effectively twice that of a single set of rolls. When the rolls are driven by fixed gears, however, the additional force depends on the accuracy with which the equipment is built and the uniformity of the wire. If for example, the wire does not pass at exactly the same distance from the axes of the driven rolls, they will attempt to drive it at different speeds and the resulting slip will waste power.

Orbital or linear wire feeders pinch the wire between rolls whose axes are inclined to the wire and which are held in a rotating cage. In some versions the cage is driven and the rolls rotate freely, in others the rolls themselves are driven and carry the cage with them. Proponents of the latter claim that less damage is done to the wire surface. The motor has a hollow armature shaft through which the wire passes. These systems can feed more than one wire diameter without changing rolls, but their geometry is such that the larger the wire, the greater the feeding speed and the lower the feeding force for a given motor speed and torque: theory would suggest that this is undesirable though in practice the solution is to provide a slightly more powerful motor. The manufacturers of orbital wire feeders claim

**6.2** Wire feeder with twin drive motors.

the ability to feed wire over longer distances than conventional types but as yet they have not made the major impact on the market that this might seem to warrant. An unchallenged benefit, however, is the straightening effect of the rolls, which is useful when working with long stickouts or in robotic applications where accurate, blind seam following is required.

Where the distance from the power source to the work cannot easily be reduced to a few metres, as for example in shipyards, repeater or push-pull wire feed units may be required. Alternatively, small, portable feeders of the 'suitcase' type, carrying a small wire spool inside, are increasingly popular. These can have a facility for selecting the welding parameters at the feeder from a pre-programmed range without the need to return to the power source.

Solid wire is used with the wire positive with respect to the workpiece and a number of welding sets have been made with the wire feeder in permanent contact with the positive side of the power source. Tubular wires, however, may sometimes need negative polarity for the best results, as when welding positionally with basic wires. To allow for changing polarity, the wire feeder and associated parts must be insulated from the rest of the equipment. Fortunately, this is now rapidly becoming standard practice for new welding sets.

Manufacturers sometimes receive complaints about wire feed units which on investigation prove to have been fitted with drive rolls from another supplier. Where these perform poorly, they are usually found to have been incorrectly hardened. As in other industrial sectors, the use of non-approved spares can be an expensive economy.

Despite recent advertising which implies that closed-loop control of wire feeders is a new concept, they have in fact long been able to provide a high degree of consistency of feeding speed. Unfortunately, few European sets have offered a facility which American welding engineers would regard as essential when using flux-cored wire, namely a wire feed speed meter. These have been standard on American equipment for many years and are most valuable in transferring welding procedures, especially for self-shielded wires, and in troubleshooting.

### Welding guns

Welding guns or torches designed for MIG/MAG welding with solid wire are generally suitable for tubular wire welding provided their current rating is adequate. Both water-cooled and air-cooled types are available. The extra inconvenience which water-cooled torches entail in the way of hoses and coolers may be offset by their smaller size and lighter weight. Torches run cooler with $CO_2$ shielding than with argon-rich mixtures, so a higher current or duty cycle may be used according to the manufacturer's recommendations. With gases containing helium, torches run a little hotter than on pure argon-$CO_2$ mixtures.

Compared with solid wires, tubular wires buckle more easily and, with certain exceptions, lack a copper coating. More thought must therefore be given to torch design to ensure smooth feeding and good electrical contact. Straight-through torches offer little mechanical resistance to the passage of the wire and were widely used with early, large-diameter, thin-walled tubes which buckled readily and could be damaged by high contact forces at the feed rolls. If the wire is straight, on the other hand, as for example when it emerges from modern packs designed for robotic applications, contact may be poor when using a straight torch. A curved or 'swan neck' type forces the wire into contact with the outer side of the curve. If feeding problems are encountered, the temptation to use a PTFE liner of the type recommended for aluminium alloys should be resisted. Properly maintained equipment with steel liners should provide good feeding if the wire is correctly lubricated.

Manufacturers supply contact tips to suit the wire sizes on the market and they would typically be bored to 0.2 mm larger than the wire diameter. However, BS 7084 and AWS A5.20 allow tolerances on wire diameter of ±0.05 mm up to 1.6 mm and up to ±0.08 mm for wires of 2 mm and upwards. Most wire manufacturers regard these as too wide and work to closer limits. When they also make equipment, they bore their contact tips, for which no standards exist, to the optimum size for their own wires. The result is that there is no absolute guarantee that one manufacturer's wire will feed through another's contact tips, though problems are rarer than they once were. The adoption of prEN 759 should reduce the difficulty at least in Europe, because tolerances will then be +0.02−0.05 mm up to 1.4 mm and +0.02−0.06 mm from 1.6 to 3.0 mm. A standard for contact tips would then be desirable.

Welding torches have to provide, where appropriate, gas shielding for the arc and while this was seen in the early days of gas-shielded welding as the main design criterion, some might claim that it has taken second place to ergonomics in modern torch design. Much work in the 1960s and 1970s showed that for the best shielding, it was necessary for the gas flow to be laminar as it emerged from the nozzle. The simplest way to achieve this is to provide a gas path within the torch that is as long as possible and parallel-sided. For semi-automatic welding a 200 or 300 mm long barrel is too cumbersome and the use of some form of diffuser at the point where the gas enters the barrel allows better results with shorter torches. Mechanised applications can offer the opportunity to use longer guns. Both gas suppliers and torch designers have been guilty of using the avoidance of porosity as the main criterion of the effectiveness of a gas shield. This ignores the fact that a deterioration of toughness in steel weld metals is very apparent as weld nitrogen levels rise from 50 to 100 ppm, whereas porosity is rarely observed until nitrogen levels rise above 200 ppm. It is difficult for users to check shielding efficiency for themselves but intelligent questioning of the supplier may produce documentary evidence.

For higher welding currents, the optimum nozzle diameter increases as shown in Fig. 4.3 in Chapter 4. Tapered nozzles help to improve access in narrow joint preparations and in that case the joint itself helps to exclude the atmosphere, so the pattern of gas flow out of the torch is less critical. In the top layer of a thick joint there is less help from the preparation and here nitrogen levels may rise slightly when tapered nozzles are used, but fortunately the top layer is hardly subject to strain ageing so toughness does not necessarily suffer. Highly tapered nozzles should be avoided when welding thin joints with open geometries.

Guns for use with self-shielded wires can be made much lighter than those for gas-shielded types since there is no need for gas shrouds or hoses. Some self-shielded wires are designed for long electrode extensions of 50 mm or more and with a standard gun this can be difficult to hold consistently; furthermore, if the wire is not perfectly straight it may not be possible to follow the joint line. The solution is to use an electrically insulated guide tube at the end of the contact tube which allows a long electrical extension but only needs a short stickout of the wire beyond the tip. These are readily available for the special guns designed for self-shielded wire.

When tubular wires were commonly used in diameters of 2.4 mm or more, the heavy guns and cables were tiring to use and counterbalanced booms were often provided to improve the welder's comfort. Fewer are sold today, but even with finer wires and lighter guns, their use is worth considering if high duty cycles are planned.

# CHAPTER 7

# Standards

The earliest standard for flux-cored wire was AWS A5.20, issued in 1969. In revised form it is still the most widely used flux-cored wire standard and it is difficult to see how any of the proposed international standards produced to date would persuade the American Welding Society and its supporters to abandon such a well-tried and user-friendly system. Some details could nevertheless benefit from clarification and updating and a new version is expected shortly.

AWS designations for tubular wires, as for MMA electrodes, start with the letter E for electrode, followed by a digit or digits indicating the tensile strength of the weld deposit in units of $10\,000\,\text{lbf/in}^2$ or 10 ksi, Fig. 7.1. The designated levels in A5.20 are actually 62 ksi ($428\,\text{N/mm}^2$) and 72 ksi ($497\,\text{N/mm}^2$), shown by the digits 6 and 7, but the values may be 1 ksi ($7\,\text{N/mm}^2$) lower for each percentage point by which the elongation increases above the specified minimum of 22%, for a maximum reduction of 2 ksi to 60 ksi ($414\,\text{N/mm}^2$) and 70 ksi ($486\,\text{N/mm}^2$) respectively.

After the code for strength in A5.20 comes a digit showing positional welding capability, 0 for the flat and horizontal positions and 1 for all positions. The standard is however subject to different interpretations in dealing with electrodes which can be used for positional welding in the smaller, but not in the larger sizes. Verification of positional capability is only required using a diameter of wire recommended by the manufacturer, so the specified requirements for an EX1- electrode can be met provided any size passes the test. On this basis, many manufacturers use the EX1- designation for all sizes of electrodes of which only the

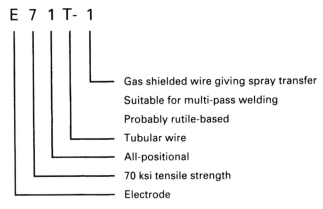

**7.1** Tubular wire designation from AWS A5.20–79.

smallest are truly all-positional. On the other hand, the standard specifically permits 'dual classification' in these circumstances so other manufacturers classify large diameters as EX0- and smaller diameters as EX1-. If, however, 'dual classification' were to have the same meaning as in A5.26 for electrogas wires, then a single diameter of a wire could have two classifications and some suppliers do this. It is hoped that the next issue of A5.20 will offer clarification. In the meantime, problems can arise with literal-minded users when a manufacturer extends downwards the range of sizes of an electrode which is then reclassified from EX0- to EX1-. Given that the requirements for the latter include all those of the former, an assurance that the product is otherwise unchanged should avoid any need for requalification.

Following the digit for positional capability, the AWS designation has the letter T for tubular wire and one or two symbols which encapsulate the nature of the product, see Table 7.1. The twelve types carry with them different testing and analytical requirements and each describes a group of products which are similar in usability characteristics and probably in core composition. Users can readily form a picture of the type of wire thus classified and often refer to them simply as a 'T-5' or a 'T-8' wire. Detailed descriptions of the classifications are contained in a non-mandatory appendix and this has caused some pain to the more purist of European standards writers for whom descriptive matter is to be viewed with suspicion, but a generation of users of the AWS specification would disagree.

*Table 7.1* Tubular wire core types in AWS A5.20

| T-1 | Rutile type for use with $CO_2$ or mixed gas for multipass welds<br>EX0T-1 for downhand only<br>EX1T-1 for all positions |
|---|---|
| T-2 | Rutile type for single pass welding, downhand only with $CO_2$ |
| T-3 | Self-shielded type for high speed on thin material<br>spray transfer on DC+ |
| T-4 | Self-shielded type for high deposition rates, single or multi-pass<br>globular transfer on DC+ |
| T-5 | Basic type for use with $CO_2$ or mixed gas<br>EX1-T5 types can be used positionally |
| T-6 | Self-shielded, giving 'spray transfer' on DC+<br>good toughness in multipass welds |
| T-7 | Self-shielded, all-positional on DC− |
| T-8 | Self-shielded, all-positional on DC−good toughness |
| T-10 | As T-3 but DC−, said to be suitable for any thickness but only tested<br>for 6.4 mm |
| T-11 | As T-7 but with 'spray transfer' |
| T-G | Any other type |

The T-1 classification describes one of the oldest types of flux-cored wire. Although the word 'rutile' did not appear in the first issue of A5.20, the description now says that most electrodes in this group have a rutile slag. Electrodes are classified with $CO_2$ shielding gas but the use of argon-$CO_2$ is now mentioned in the Appendix: in this respect the AWS lags behind current European practice and standards which admit rutile wires specially formulated for use with argon-based mixtures. Despite the efforts of American manufacturers to create a distinction between the older E70T-1 wires and the newer E71T-1 types, both still appear as T-1. The European division of rutile electrodes into two classes has perhaps more to do with establishing a new image for the product than with any real technical distinction.

T-2 wires are little seen in Europe and those that are found are imports. They have higher deoxidant levels compared to the T-1 types which makes them more suitable for single-pass fillet welding on oxidised plate or, it is claimed, for high dilution welding with one pass from each side. As discussed in Chapter 4, the European view is that current T-1 types make the T-2 wires superfluous.

When the AWS classification was introduced, all the wires from T-3 onwards could be used without shielding gas, but when T-5 wires ceased to be recommended for gasless use this left the types interspersed, a small untidiness which current standards in Europe avoid. The T-3 wires are intended for gasless welding of thin sheet at high speeds. As described in Chapter 2, this type of wire may be based on rutile, since the relatively low ductility and toughness which result do not necessarily cause concern for thin sections. As in the case of the T-2 wires, the standard only requires a transverse tensile test and a longitudinal guided bend by way of checks on mechanical properties.

The T-4 group of wires are the earliest of the purely self-shielded types. Although the AWS classification does not specify the slag components, they are usually based on calcium fluoride with aluminium as a deoxidant and nitride-former. This allows the welds to develop a respectable level of tensile ductility but notch toughness is not their forte and is not tested by the standard. These wires are in widespread use for general fabrication of structures and components whose toughness is not critical, especially in countries where shielding gas is not readily available. They are used with electrode positive, giving a rather globular transfer but a slightly softer arc and lower spatter than they would display with electrode negative.

As already mentioned, T-5 wires are something of an anomaly in the AWS specification, which pays the penalty for being first in the field at a time when the basic flux-cored wires were used either with or without shielding gas. It is still possible, indeed, to run many modern T-5 wires without gas but today's users would not tolerate the long stickout needed to prevent porosity nor the lack of welder appeal. Present T-5 wires occupy a niche where above all an exceptional combination of strength and toughness is sought, together with the lowest hydrogen levels and the best tolerance to contamination by rust, scale, grease or paint. In recognition of this the AWS requires T-5 wire to undergo impact testing at $-20\,°F$ ($-29\,°C$). The group is described as having a lime-fluorspar based slag.

The T-6 classification has been transformed since the first issue of A5.20, which described T-6 wires as 'similar to those of the T-5 classification, but . . . designed for use without an externally applied shielding gas'. This implied a globular transfer, low penetration and probably less than perfect slag removal. In contrast, A5.20 now describes T-6 wires as giving spray transfer, deep penetration and excellent slag removal. The wires have been developed away from their T-5 origins and, making extensive use of synthetic materials in the core, now offer good productivity and welder appeal while retaining good toughness. They have to meet the same impact requirement as the T-5 wires.

Relatively little change in formulation is needed to make a T-4 wire into a T-7 type and users of the first issue of A5–20 were given no guidance as to their different applications. All was made clear in the 1979 edition, when it was revealed that the smaller diameters of the T-7 wires may be used for positional welding. The use of electrode negative polarity gives a finer droplet size to assist in this. Larger diameters have similar uses to those of the T-4 types but in some cases give more penetration. The polarities assigned to the T-4 and T-7 wire when not used positionally are somewhat arbitrary: one of the largest T-4 users prefers to work with electrode negative and some T-7 wires give good downhand results with electrode positive.

No wires were described as suitable for positional welding in A5.20–69, mainly because the small diameter wires available today were not produced then, but the T-8 wires are now regarded as an all-positional type combining good weldability and toughness. As described in Chapter 2, this is achieved through the use of lithium and barium compounds. Like T-5 and T-6 wires, the T-8 class has an impact requirement of 20 ft. lbf at −20 °F (27 J at −29 °C).

A5.20 has not yet listed any T-9 wires but T-10 and T-11 types appeared for the first time in the 1979 issue. T-10 wires resemble those of the T-3 group in being designed for high-speed welding, but operate with electrode negative instead of positive. Ductility is likely to be rather better for the T-3 wires but given that neither type is tested by more than a longitudinal guided bend, it is doubtful what the user gains by the separate listing of the two types. Similarly, it could be argued that since the requirements for T-7 and T-11 wires are identical in the body of the standard, little purpose is served by having two classes. The fact that T-11 wires have improved metal transfer and can operate positionally is not mentioned in the standard itself and from a European viewpoint could have been acknowledged by a simple note under the T-7 description.

Wires not classified under any of the above categories may be listed as T-G if they are suitable for multi-pass welding or T-GS if they are for single pass welding. The most significant class of products which hitherto have had to be listed as T-G are the metal-cored wires, which were not specifically described in A5.20-79. The AWS has now addressed this by including them, not in the flux-cored wire standard, but in A.5.18, which previously dealt only with solid wires. In A.5-18-93, metal-cored or 'composite' wires are designated E70C- (c.f. ER70S- for solid wires), followed by the digit 3 or 6, denoting a Charpy requirement of 27J at $-18\,°C$ or $-29\,°C$ respectively. Alternatively, the letter G may be used where no chemical analysis is specified and the toughness is agreed between supplier and purchaser. There follows a letter C designating use under $CO_2$ or M for argon-$CO_2$ mixtures, a nuclear designator N where stringent control of copper, vanadium and phosphorus is claimed and, for the first time, an optional hydrogen designator allowing manufacturers to claim a diffusible hydrogen level of 4, 8 or 16 ml/100 g.

The new A5.18 specification will remove the anomaly which has led wire suppliers until now to describe similar tubular wires variously as E70T-1, E70T-5 or E70T-G to the confusion of users. However, many users will feel that an opportunity has been missed to incorporate all tubular wires in a single standard, as has been the practice in Europe.

European criticism of American standards may justly be seen as presumptuous but two pleas of mitigation are offered in respect of the above comments. First, the AWS specifications have earned themselves such a following in Europe that no European welding manufacturer can ignore them. Secondly, new European standards are in course of preparation and if these are to embody the best of current practice from around the world, a constructive analysis of what is on offer elsewhere is needed.

Such is the pace of European standardisation indeed that it is only necessary to touch briefly on present European national standards insofar as they represent steps on the way to eventual European standards. These will automatically supersede national standards as soon as they are adopted by CEN, the European Standardisation Committee. The British standard for tubular welding wire, BS 7084, was introduced in 1989 and anticipated in large measure the format of a wave of European standards for welding consumables which are expected to start with the publication of EN 440, dealing with solid wires for gas-shielded welding,

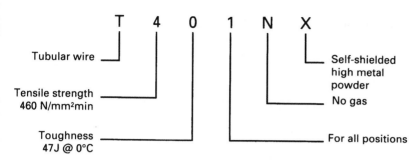

**7.2** Tubular wire designation from BS 7084.

in 1994. The tubular wire standard, EN 758, will not be far behind.

Starting with the letter T for tubular wire, there follow in BS 7084 two codings representing strength and toughness respectively. The minimum tensile strength levels 430 and $510\,N/mm^2$, designated 4 and 5 respectively, correspond roughly to those represented by 6 and 7 in AWS specifications. Impact toughness is given by a digit showing the temperature at which 47 J is achieved: 2 for −20, 3 for −30 °C and so on. European drafts almost certain to be adopted from 1994 use the same coding for toughness but use two digits to represent yield strength instead of one to show tensile strength. The strength and toughness designations will be common to European standards for welding consumables for all processes. A greater number of grades of strength appear in the European drafts than in most existing national standards and CEN has resisted alignment with the latest ISO draft for MMA electrodes.

The popularity of the descriptive element in AWS standards led to the inclusion of a similar section in BS 7084, Fig. 7.2. In this case two letters appear, the first G or N indicating Gas or No gas shielding required. The second letter indicates the type of slag system or operability.

The first descriptive letter in BS 7084 and in prEN 758 is R, describing a rutile electrode with slow-freezing slag, suitable for welding in the flat and horizontal-vertical positions, see Table 7.2 and Fig. 7.3. Both standards then use the letter P for positional rutile types, thereby creating a distinction not present in the descriptive part of the AWS specification.

*Table 7.2* Tubular wire core types in BS 7084 and prEN 758

| BS 7084 description | Letter | prEN 758 description |
| --- | --- | --- |
| For spray transfer, flat and HV | R | Rutile base, slow freezing slag |
| For spray transfer in all positions | P | Rutile base, fast freezing slag |
| Basic flux | B | Basic slag |
| Metal-cored | M | Metal powder core |
| For high deposition rate, flat and HV | U | — |
| For single run, high speed, flat and HV | V | Rutile or basic/fluoride |
| For multi-run, flat and HV | W | Basic/fluoride, slow freezing slag |
| High metal powder/low flux: all pos. | X | — |
| High flux/low metal powder: all pos. | Y | Basic/fluoride, fast freezing slag |
| Other types | S | — |
| — | Z | Other types |

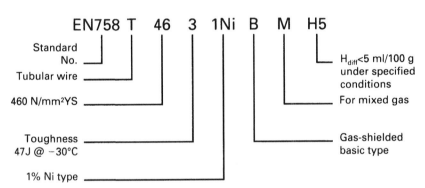

**7.3** Tubular wire designation from prEN 758.

BS 7084 and prEN 758 use the letter B to denote wires with basic fluxes which correspond to the T-5 group of the American system. Last of the gas-shielded wires is the metal-cored type M, defined as containing less than 1% non-metallic materials as a percentage of wire weight. All these gas-shielded types will be familiar to users for whom 'rutile', 'basic' and 'metal-cored' are terms in daily use to describe tubular wires.

In arriving at a system for classifying and describing self-shielded wires, European standards writers found themselves in a quandary. The wires are seen as transatlantic in origin and in a market dominated by one or two large suppliers, the AWS quite legitimately wrote their classification round particular commercial products which would be familiar to users. European users are much less familiar with the products, but core compositions are not generally in the public domain, so to classify products by their core composition is not very helpful either. The solution adopted for BS 7084 was to take at random the letters U, V, W, X and Y to denote wires corresponding approximately to the AWS types T-4, T-3/T-10, T-6, T-7/T-11 and T-8 respectively. No attempt was made to go into the slag chemistry but wires were described mainly in terms of operating characteristics.

In the CEN system, a further simplification took place to leave only three self-shielded categories: V, W and Y. The first of these comprises all the wires designed for single or two-pass welding only. They include rutile and fluoride-based products for down-hand welding as well as some all-positional wires currently listed as T-GS under the AWS classification.

Electrodes classed 'W' in prEN 758 include, as well as the T-6 wires for which W chiefly stands in BS 7084, all the other down-hand types suitable for multipass welding: in AWS terms, the T-4 and some T-7 wires. CEN decided that it was untidy to describe MMA slags by their chemistry and not those of flux-cored wires, and the fact that few people knew what went into the latter has resulted in the rather uninformative description 'basic/fluoride' for this class, which now embraces both the older wires with fluorspar as the main slag former and newer wires which contain much smaller amounts of fluorides in general and fluorspar in particular.

The final specific classification in EN 758 is type Y, which includes all the self-shielded wires suitable for single and multi-pass, all-position welding. Once more, the basic/fluoride slag description masks a progression from the simple lime-fluorspar systems of early wires to modern types based on oxides of lithium, barium and magnesium and from unspecified toughness to adequate properties at $-20\,^{\circ}C$ and below.

As in many other standards, an undefined type designated 'S' or 'Z' respectively is included in BS 7084 and EN 758, corresponding to 'T-G' and 'T-GS' in the AWS document. This allows newly developed products to be classified in terms of their mechanical properties within the standard until such time as they are sufficiently well established to be given a descriptive designation in their own right.

In EN 758, the letter showing the gas shielding requirement follows that describing the slag and offers the option of classification either with $CO_2$, denoted C, or with argon mixtures, M, in addition to N for no gas. BS 7084 simply requires the recommended gas to be stated in the labelling.

EN 758 and other European welding consumable standards adopt the welding position digit from the International Standard ISO 2560. They differ from this and from many national standards, however, in demanding proof of positional capability in the form of fillet weld tests, which must produce beads meeting specified criteria for throat thickness, convexity, concavity and equality of leg length.

The last symbol in BS 7084 and EN 758 allows the manufacturer optionally to claim controlled weld hydrogen. In BS 7084 the single letter H indicates a diffusible hydrogen level below 15 ml/100 g deposited metal. The manufacturer states in his literature whether the actual level claimed is 5, 10 or 15 ml/100 g and what restrictions need to be placed on welding parameters to achieve this result. EN 758 uses the codings H5, H10 and H15 but the user still has to refer to the manufacturer's literature to find out how to arrive at the specified level. As described in Chapter 4, testing for diffusible hydrogen is fraught with difficulty and users should certainly be aware of the effect that testing conditions can have and should read the information on parameter restrictions with care.

AWS A5.20 and BS 7084 do not include alloyed wires within their scope. The CEN approach, however, is to group welding consumables according to the types of steel they are designed to weld. In the case of EN 758, these are 'non-alloy and fine grain steels'. Since non-alloyed steels are often welded with low-alloy consumables, containing for example nickel or molybdenum, these are included in EN 758. To simplify matters where no alloying is used, no symbol for chemistry appears when the manganese content is less than 2%, nickel is less than 0.5% and molybdenum less than 0.2%.

Where alloying is intentionally added, EN 758 gives a table of designations to be used for grades up to 3% Ni and 0.5% Mo. Higher alloy contents are not thought necessary in a standard which only includes yield strength levels up to $500\,N/mm^2$. Alloy designations are straightforward, for example 2 Ni or 1 NiMo, and are placed between the toughness symbol and the symbol showing core type. Further standards for tubular wires for creep-resisting and stainless steels are planned for issue at the end of 1995 but no work item yet exists for a standard on tubular wires for high strength steels.

It should be mentioned here that the International Standards Organisation does produce standards for welding consumables and in the 1970s, many countries outside the USA adopted ISO 2560 as the basis of their national standards for MMA electrodes. Drafting is in practice delegated to the International Institute of Welding, who are currently working on two standards for tubular cored wires. The first deals with wires for welding carbon-manganese steel and the second with wires for low alloy steel. As might be expected, the drafts appear as hybrids between European and American standards and, as is sometimes the case with hybrids, they do not inherit all the best features from each parent. However, they are workmanlike documents and if they had seen the light of day ten years earlier might have been widely adopted. As it is, they come too late to influence the European standards and the AWS has no history of adopting ISO standards, so the future of the drafts must remain in doubt.

Low alloy tubular wires are covered in AWS A5.29. This includes weld metals with yield strengths up to $750\,N/mm^2$ and as well as the Mn-Mo-Ni grades for structural and pressure vessel steels, the creep-resisting grades up to $2\frac{1}{4}$ Cr 1 Mo. The first part of the A5.29 designations resemble those of A5.20 except that strengths go up to E12 for 120 ksi, but after the code for core type there is another which gives the deposit chemistry. These are encrypted as A1 for 0.5 Mo, B1 to B3 for the Cr-Mo grades, D1 to D3 for the Mn-Mo grades, K1 to K7 for the Ni-Mo and Ni-Cr-Mo types and W for the Ni-Cr-Cu weathering grade. The 1, 2 and 3% Ni grades are written in clear as Ni1, Ni2 and Ni3 respectively. Appropriate minimum toughness requirements are laid down at temperatures down to $-100\,°F$ $(-73\,°C)$ for the Ni3 types. For the creep-resisting grades and certain other types likely to be subject to post-weld heat treatment in practice, mechanical testing of the welds for classification is carried out in the PWHT condition.

Stainless flux-cored wires are specified in AWS A5.22 which, in the absence of comparable standards in Europe, has achieved wide currency. It is based on the familiar American stainless steel designations in which the 300 series are austenitic and the 400 series ferritic and martensitic. While this was a simple and compact system in its original form, new grades have been added by means of suffixes showing additional elements, as in E308Mo or E410NiMo. This is starting to be cumbersome and has led to some inconsistencies of presentation: for example, a niobium-stabilised E308 becomes an E347 but a stabilised E309 becomes E309Cb. It seems likely that CEN will use the more representational coding system in which 308Mo becomes 20 10 3. Both systems use an -L suffix to denote low carbon.

After the coding for chemistry in A5.22 come the symbols T-1, T-2, T-3 or T-G. These do not, as in A5.20, describe the flux type but the shielding gas used for classification. For T-1 this is $CO_2$, for T-2 $Ar+2\%O_2$, for T-3 no gas is used and for T-G the gas is not specified. The choice of gases seems rather dated now, when most stainless flux-cored wires sold in Europe are used with 75–80% Ar, 20–25% $CO_2$. This is significant because the maximum deposited metal carbon content for the -L grades is 0.03% for the T-2 wires and 0.04% for the T-1 types. Users for whom this is important have therefore to seek additional guidance when they are using argon-$CO_2$ mixtures.

## Gas standards

The UK has not until now had a British Standard for shielding gases, but the European draft prEN 439 is likely to be approved shortly. The format is that of the DIN standard 32 526, although there are significant changes of detail. Impurity levels are calculated on a different basis and it is not immediately apparent that they may be substantially higher. Gases are classified by oxidising potential and those of interest to tubular wire users are those designated M for mixed gas and C for $CO_2$. Mixed gases are based on argon and are grouped into three classes M1, M2 and M3 in order of increasing oxidising capacity. Each class is then further subdivided with a second digit.

At this point users need to be cautious, because although the new designations may look the same as in DIN 32 526, the mixtures specified may be different. For example, the M11 type in the DIN standard is $Ar-2\%O_2$, but M11 in prEN 439 also contains up to 5% hydrogen. The direct equivalent in prEN 439 is M12. The brand of $Ar-20\%CO_2$ most popularly used in the UK contains about 2% oxygen and was often, though not strictly correctly, described as M21, the mixed gas most common in continental Europe: it now becomes M24. As welding procedures are updated, it will be important to ensure that EN 439 gas mixtures are entered by reference to the standard rather than by simply copying the previous DIN designation.

# CHAPTER 8

# Applications and consumable selection

It has been said elsewhere that there are today no 'no-go areas' for flux-cored wire. Objections to the process previously raised on the grounds of quality are out of date and the range of products available is such that almost any weldable material can be tackled. But welding engineers are no less conservative than any other professionals and the spread of the process has been principally by example. Although slow to invade new industries initially, after the first successful application tubular wire welding is driven through an industry by competitive pressures. A short survey of the international scene shows how this is still happening today.

The USA was the commercial birthplace of flux-cored wire although others might lay claim to its paternity. Bridges and high-rise buildings were among the first applications for the process. Europe has built some bridges with flux-cored wire but general structural work lags far behind the USA. Particular examples such as the Heathrow maintenance hangar or the new terminal at Stansted Airport stand out among a mass of bolted structures. For welding on site, and especially above ground level, self-shielded wires have been much used in the USA, but where shop prefabrication is possible, gas-shielded wires can improve productivity.

Offshore fabrication became, in the 1970s, a showcase for welding with flux-cored wire. The deep water of the North Sea required unusually large platforms with section thicknesses of 50 mm and more, while the low ambient temperatures meant that extra attention had to be paid to the possibility of brittle fracture. The recently developed CTOD test became the standard for off-shore steels including weld metals. The heavy joints and the large volume of welding needed put a premium on high productivity and flux-cored wires were an obvious choice for the work if the stringent toughness criteria could be met. Self-shielded wires were used first,[43] because welding at the exposed coastal sites was often poorly protected from wind, even inside the shops. Later, with better exclusion of draughts, a new generation of high-toughness, gas-shielded rutile wires has gradually supplanted the self-shielded types.[50] Large tonnages of wire, mainly of the E81T-1Ni1 grade, are used in offshore fabrication yards in Britain, Norway, Finland and Spain in particular with smaller yards and sub-contractors through the whole of Europe. Where clients have waived CTOD requirements for welds for which adequate Charpy toughness at $-60\,°C$ can be demonstrated, some fabricators have opted for E81T-1Ni2 wires.

The shipbuilding industry is probably the largest tonnage user of welding consumables and the Japanese shipbuilding industry rose to world dominance at a time when it moved towards the use of a high proportion of flux-cored wire welding, initially with E70T-1 wires. Europe was slow at the outset to follow the Japanese lead but Finland pioneered the move to tubular wires in the early 1980s.[51] Metal-cored wires were the first to allow welding over prefabrication primers and for welding icebreakers, 2% Ni wires gave the necessary toughness down to $-60\,°C$. An alternative approach to welding over primers, used successfully but limited by the capital investment needed, was pulsed arc welding with a basic wire. This allowed high deposition rates in HV fillets but was not used positionally.

The Japanese strategy for welding over primers was firstly to standardise on the relatively weldable low-zinc silicate type and then to use metal-cored wires designed, like most wires for use in Japan, to run under $CO_2$. The same primer technology was adopted in Europe, slowly at first in the UK, but metal-cored wires have hardly been used under $CO_2$ here, despite the availability of competitive products. An alternative and more recent Japanese approach has been the use of wires with about half the normal volume of rutile-based slag, designed to allow the gases evolved during welding over primer to escape.[52] Some European yards have even been able to weld vertically over primer, using E71T-1 wires whose slags are not the stiffest available, with Ar-$CO_2$ shielding. For horizontal-vertical welding, metal-cored wires are

**8.1** The world's oldest all-welded ship was launched in 1920. Shipbuilding is the largest user of flux-cored wires today.

still difficult to beat in terms of productivity. Some current primers allow the use of argon mixtures for shielding, but if future needs for better corrosion protection or solvent reduction lead to different primers, $CO_2$ shielding or even new, more complex gas mixtures may give the extra tolerance required.

Much welding in shipyards can be done using submerged arc welding (SAW), and with its low environmental impact and high productivity this is not a process that gas-shielded welding should be trying to replace. However, the use of tubular, and in some cases even flux-cored wire under a submerged arc flux has aroused considerable interest in the shipbuilding industry and offers some of the advantages of both processes. Productivity can be improved in two ways compared with solid wire SAW. In the first place, deposition rates can be higher with a tubular wire. Although it was seen in Chapter 2 that this argument does not always hold for small diameter metal-cored wires under gas, various factors combine to tilt the balance in favour of tubular wires for the SAW process. Secondly, the use of small amounts of fluorides in the core of tubular wires can increase their tolerance to welding over primers, as measured by maximum welding speed, without the need to resort to a highly basic flux. Increases in welding speed in shipbuilding applications of 20% for fillets on primed plate and 30% for butt welds are possible for tubular wires in comparison with solid wires.

Tubular wires can also produce submerged arc welds with improved toughness where considerations of weldability or hygroscopicity make highly basic fluxes less than a preferred option. Not only can the addition of basic minerals to the core allow users the possibility of combining basic flux properties with acid flux usability, but further benefits arise from being able to add other elements such as strong deoxidants which would not survive the high temperatures involved in flux manufacture. Although all this was first mooted many years ago, it is only recently that economic and technical pressures have overcome the natural conservatism of shipbuilders and that tubular wires have been used in the yards.

Submarine construction is a special case of shipbuilding where flux-cored wires have started to be used in Europe but are not yet accepted in the United States. The interest here is in the type of steel to be welded, generally quenched and tempered types of 550, 690 and even $900\,N/mm^2$ minimum yield strength. At the $550\,N/mm^2$ level, rutile wires can comfortably be used. At higher levels, only basic wires are at present approved though new rutile wires offering very low hydrogen contents may be a possibility for the future.

The industry which more than any other brought welding with flux-cored wires to Europe was the manufacture of earthmoving equipment.[53] American companies setting up manufacturing plants in Europe saw that they could achieve better productivity than local manufacturers by bringing in technology they already had at home, and that initially meant 2.4 mm E70T-1

**8.2** Submarines have pioneered the use of quenched and tempered high strength steels.

wires under $CO_2$ or E70T-6 self-shielded wire. As shown in Chapter 3, these could give very high deposition rates and users, aware of the attendant high fume emission rates, took care to provide adequate fume extraction. At first, European consumable manufacturers were unable to provide equivalent wires with an acceptable guarantee of quality, but by the mid-1970s they were supplying the industry with competitive products. Today, E70T-1 wires are still used but self-shielded types are less seen. As automation and robotics are increasingly used, metal-cored wires are starting to replace flux-cored types.

Moving down the size scale from earth-moving equipment to trucks, trailers and off-road vehicles, self-shielded wires have again been used to a surprising extent which reflects American influence in these industries, but in Europe metal-cored wires are gaining ascendancy. Competition from solid wires is strong, but in this field as in that of automotive components in general, accurate costing and measurement of productivity can justify the use of tubular wires.

Pipeline welding is a sector where productivity is almost the sole factor in the choice of consumables given that quality and safety are assured. This has resulted in the development of cellulosic electrodes and the technique of stovepipe welding, a combination which has proved difficult to beat. The linear speed of rooting would typically be 0.25–0.3 m/min (10–12 in/min) and deposition rates above 2 kg/h are possible for the filling passes. The good penetration allows cellulosic electrodes to be used in the vertical-down direction without lack of fusion. However, self-shielded tubular wires are now available specially formulated for good penetration and control, and these closely match the MMA speeds and deposition rates in semi-automatic use. A 1358 km (844 mile) line of 24 and 30 in (610–762 mm) pipe was laid in Argentina by this process in 1985.[54]

More significant productivity improvements come when mechanised welding is used. Self-shielded wires operating in the vertical-down direction can reach deposition rates around 3.3 kg/h with acceptable slag control and penetration. Rutile gas-shielded wires have so far not proved successful in the vertical-down direction as their penetration has been marginal. It may be that in the future, specially developed rutile wires will succeed in this application. Meanwhile, standard rutile wires are used vertically up at

| | | | | | | | |
|---|---|---|---|---|---|---|---|
| Pipe welding using the CRC process with E 81T-1 Ni1 wire | | | | | | | |
| Pass No | Root | Hotpass | Fill 1 | Fill 2 | Fill 3–4 | Cap 1 | Cap 2 |
| Process | MMA | MMA | FCAW | FCAW | FCAW | FCAW | FCAW |
| Travel direction | Downhill | Downhill | Downhill | Uphill | Uphill | Uphill | Uphill |
| Electrode diameter, mm | 4 | 4 | 1.4 | 1.4 | 1.4 | 1.4 | 1.4 |
| Wire feed speed, m/min | | | 6.6 | 7.6 | 7.6 | 5.3 | 5.3 |
| Shielding gas | | | 80/20 | 80/20 | 80/20 | 80/20 | 80/20 |
| Gas flow rate, l/min | | | 40 | 40 | 40 | 40 | 40 |
| Wire extension, mm | | | 12 | 12 | 12 | 12 | 12 |
| Current, A | As req | As req | 240–260 | 280–300 | 280–300 | 220–240 | 220–240 |
| Voltage, V | | | 22.5 | 23 | 23 | 22 | 22 |
| Travel speed, m/min | As req | As req | 0.28 | 0.15 | 0.20 | 0.18 | 0.18 |

60 mm pipe × 21 mm wall thickness

**8.3** Mechanised pipe welding with tubular wire.

rather lower deposition rates. District heating pipes of 24–40 in (610–1016 mm) diameter and totalling some 20 km in length were laid in Copenhagen and Berlin using this process during the 1980s and a very low repair rate was claimed.[55]

An automatic pipewelding method developed in the former USSR for use with flux-cored wire, the 'Styk' system,[56] may be regarded as intermediate between conventional flux-cored arc welding and electrogas welding. High currents of 300–400 A are employed using a 2.4 mm wire and water-cooled shoes travel round the outside of the pipe to contain and control the weld pool. The cross-section of each layer may be up to 150–200 mm$^2$ so very high toughness is difficult to achieve, but productivity is high on the pipelines of more than 1 m (40 in) diameter for which the process was designed.

Structural fabrication, ships and heavy vehicles have been the areas where largest amounts of flux-cored wire have traditionally been used but until now, pressure vessels and pipework for power generation and chemical process plant have not made full use of the process. This is partly because such plant is less tolerant to defects than many other welded structures and safety-conscious designers were not in the past convinced of flux-cored wires' ability to produce defect-free welds. It is also because many of the alloy types needed to weld pressure vessels have not been as freely available as tubular wires as they have in MMA electrode form. Today, the range of tubular wires should allow most pressure vessel steels to be welded. For the first time, some Cr-Mo wires are being used in the manufacture of power generation equipment. Process plant makers have been slow to move to tubular wires, but modern NDT techniques should reassure them that quality will not be put at risk. For plant such as hydrocrackers, toughness is often more critical than very high creep rupture strength and it has been shown that embrittlement following long or repeated stress-relief treatment depends critically on the levels of residual impurity elements in the weld metal. Solid low-alloy wires have for many years been made from steel with a high scrap content and it has proved difficult and expensive to achieve the desired residual levels by this route. Tubular wires made from low-residual, mild steel strip and filled with high purity powder have long been a viable alternative to solid wires for submerged arc welding of Cr-Mo steels and will be increasingly seen in this application in the future.

Among the more highly alloyed ferritic steels, armour plate represents a considerable tonnage and lends itself ideally to welding with tubular wire. Differences in philosophy on suitable compositions for welding armour go back to World War II: in 1943, both the USA and Germany were concluding that an 18–8 steel with added manganese, now known in its American version as a Type 307, was best. However the UK has continued to use a 20Cr-9Ni-3Mo filler material while a 29Cr-9Ni or Type 312 filler is often used in equivalent civilian applications. Heavy tanks consume large amounts of weld metal in their manufacture so to achieve satisfactory productivity, MMA electrodes of up to 10 mm

**8.4** A stainless steel rotary dryer hood including section thicknesses up to 48 mm welded with rutile flux-cored wire.

diameter have been regularly used. Today, flux-cored wires can provide similar deposition rates and higher duty cycles. Some users have favoured self-shielded wires and guns having integral fume extraction to remove the large amounts of fume generated at these high deposition rates. Alternatively, gas-shielded wires produce less fume and although extractor torches need to be used with caution in gas-shielded welding of ferritic steels, fewer problems with nitrogen contamination arise when the filler is austenitic. Most welding of armour is carried out in the flat position, but some small diameter all-positional tubular wires are available. For robotic installations, metal-cored wires can give exceptional deposition rates but they are at their best when used with helium gas mixtures and the high levels of radiation and ozone generated make them uncomfortable to use semi-automatically.

Stainless flux-cored wires have been available for almost as long as the process itself but their use has been limited by the absence until recently of small diameter products with good welding characteristics. As recently as 1987, a Welding Institute study[57] reported tests with stainless tubular wires in which only 2.4 mm diameter wires were used. Today, 1.2 mm is easily the most popular size for joining work and 0.9 and even 0.8 mm wires are marketed. These are used to weld chemical process plant where accurate control of weld composition is essential, and problems on that account seem to be virtually unknown. Tubular wires are also used to weld food processing plant, where stresses are low and the environment not exceptionally corrosive, but where the weld finish has to be particularly good so that the surface cannot become contaminated with trapped food particles. Modern rutile flux-cored wires produce the finish required.

# CHAPTER 9

# Health and safety

Health and safety precautions when using flux-cored wire are not very different from those demanded by other arc welding processes and users who have established safe working practices with these will find little difficulty in making the process change. The Health and Safety Executive, the Welding Manufacturers' Association and the American Welding Society among others publish information on safety in welding.[58-60]

Electrical safety must be considered for all arc processes and the user's responsibility is to ensure that all equipment, including cables, terminals and switchgear, is of adequate capacity and correctly wired in accordance with IEE or other appropriate regulations. Equipment, especially cables, should also be regularly checked for damage. Terminals and live components must be adequately protected and isolation switches must be readily accessible. The relatively low open circuit voltage required by tubular wire as compared with certain types of MMA electrode is helpful from the point of view of safety and modern auxiliary equipment such as wire feeders operates on only 24 or 42 V, compared with 110 V in the past. During welding, wire on the reel will be at the welding potential and the use of non-insulated wire baskets instead of plastic spools to hold the wire increases the volume of material at that potential. Extra precautions might then be needed in confined spaces under damp conditions which could render even low voltages dangerous. When welding above ground level, even a small electric shock could cause a fall and injury.

All arc welding processes generate large amounts of heat and detailed rules are laid down for welding on vessels which contain or have contained flammable materials. A greater or lesser amount of spatter is produced by welding so protective clothing should be worn and should include gloves. Discomfort or heat exhaustion may result from welding in confined spaces, especially if preheat is used.

## Radiation

Radiation is the commonest cause of discomfort in welding. Ultra violet rays of wavelength less than 400 nm are produced by the arc and welders are protected by headshields fitted with filters to remove both this and infra red radiation. BS 679 sets out specifications for filter glasses of different density and recommends suitable grades for welding operations. For flux-cored arc welding, not yet specifically covered by the standard, grades appropriate to MAG welding should normally be chosen: that is Grade 12EW (for Electric Welding) below 175 A and Grade 13EW at higher currents. Lighter grades are sometimes used where high fume levels are present, as for example with self-shielded products. In practice, though, it is not usually the welder himself but other workers and passers-by who are most at risk from radiation. Even short exposure to direct arc light can give rise to a painful condition known as 'arc eye' which may only become apparent some time after exposure. The eyes water and feel as if full of grit. Because flux-cored arc welding is often used at higher currents than other open-arc processes, radiation levels are high and it is particularly important to screen welding areas to protect other workers and the public. Welders are of course taught to warn others in the immediate vicinity before striking an arc. Burning of skin can occur up to several feet from a welding arc and welders and operators of automatic and mechanised welding equipment need to be protected even if they are not handling the torch directly. Welders should ensure that they present no areas of unprotected skin to the arc: even the back of the neck is vulnerable if the reflectivity of the surroundings is high. At the very high currents sometimes used in mechanised welding, normal overalls do not guarantee protection against burning and special materials may be needed if the operator is close to the arc.

## Fumes

Although there are very few authenticated cases of welders being harmed by fumes, a great deal of attention is rightly paid to these because of the possibility of long-term or cumulative effects which might not be apparent at the time of welding. Fumes are divided for classification into gaseous and particulate components and are subject to limits imposed, in the UK, by the Health and Safety Commission.[61] These limits refer to samples taken in the operator's breathing zone. There is no simple way, despite proposals from Sweden and elsewhere, of relating such measurements to consumable fume emission rates with any degree of quantitative accuracy. However, it is important to know the composition of particulate and gaseous fume in order to establish whether special precautions may be necessary.

A long term exposure limit (8 h time-weighted average) of $5 \, \mathrm{mg/m^3}$ applies to general welding fume as to other dust found in industrial workshops, with a total respirable limit of $10 \, \mathrm{mg/m^3}$. The main constituent of fume from mild steel consumables is iron oxide, which if inhaled may settle in the lungs and be detectable on X-rays. This condition is known as siderosis or 'welder's lung' and is believed to be benign and not to lead to other complications such as pulmonary fibrosis: it slowly clears when welding stops.

Sufficient fume extraction must therefore be provided for all welding processes so that the welder is not subjected to more than $5 \, \mathrm{mg/m^3}$ of total particulate fume. In most cases this can be achieved by the use of movable extractor hoods connected either to a central extraction system or to local extractors. Fish-tail extractors which can be effective in MMA welding may disturb the shielding if placed too close to the arc in gas-shielded welding, yet are often not powerful enough to deal with the heavier fume from welding with flux-cored wire. On-gun extraction removes fume well in semi-automatic welding and is particularly suited to the self-shielded process, but again care needs to be exercised if the process is gas-shielded and weld nitrogen contents are critical.

To test whether additional measures are needed because of other components of the fume, the proportion of each component is multiplied by 5 and checked against the exposure limit for that material. For example, the exposure limit for manganese is $1 \, mg/m^3$ so any fume containing more than 20% of manganese must be controlled to a lower level than the $5 \, mg/m^3$ allowed for general welding fume. In the UK, this information is included in manufacturers' fume data sheets which should be available at the point of sale of the consumable.

Of the fume constituents which are associated with more serious problems than iron oxide, many metals can give rise to a short-term condition known as metal fume fever. The commonest cause is zinc, which is not generally a component of welding consumables but is evaporated and oxidised when galvanised or painted plate is cut or welded. Other metals such as manganese, nickel, chromium and copper, which may all be evolved from welding consumables themselves, can also cause metal fume fever. These effects are not thought to be cumulative and patients recover in 24 to 48 hours.

The possibility of carcinogenic material being present in welding fume was not mentioned in HSE Guidance Note MS15 of 1978[62] but has since become the main concern of occupational hygienists working with welding, although no positive epidemiological evidence for increased cancer incidence in welders has been produced. In particular, hexavalent chromium has been recognised as a potential carcinogen[63] and the Occupational Exposure Limit (OEL) for this has been set at $0.05 \, mg/m^3$. In most stainless steel welding, this will require that the welders wear some form of personal protection to supplement local extraction systems. Air-fed helmets into which fresh air is pumped from a filtered inlet on the welder's back are increasingly used. Many consumable manufacturers now provide figures for both $Cr^{VI}$ and total Cr in welding fume. New stainless steel consumables are also starting to appear which have been formulated to give low $Cr^{VI}$ levels and these may be helpful in countries such as the UK where national codes make the distinction between total and hexavalent chromium. Some other countries, however, regard the difficulty of analysing for $Cr^{VI}$ and of separating out its effects on health as sufficient reason to impose the lower limit on total chromium in fume.

Very recently, nickel in welding fume has become a cause for concern although again there is at present no epidemiological evidence of its having harmed welders. Research is being carried out in a number of centres, but since nickel in large amounts is most commonly found in stainless steel where there is also chromium, specific precautions to deal with nickel should only be necessary in welding nickel alloys.

Barium was not mentioned specifically in the 1978 Guidance Note but its soluble compounds are registered poisons, affecting the heart muscle among others and causing cardiac arrhythmia. As pointed out in Chapter 2, the physical and electrical properties of barium and its compounds make them attractive to the formulist and almost indispensable to the developer of self-shielded wires for high toughness. The compounds used are not soluble ones: indeed, barium carbonate is so insoluble that a popular sugar refining process relied on precipitating it from a solution of barium sucrate. Discussion therefore centres on whether the barium compounds in the wire are transformed by the arc into a soluble form, or at least become so finely divided that they can be absorbed as if soluble. In the absence of clear evidence on this, it is sensible for welders to use personal protection when welding indoors as well as general fume extraction. Self-shielded wires also lend themselves to the use of extractor torches since there is no external gas shield to disrupt. The heavy deposit formed alongside self-shielded welds may be rich in barium and care must be taken not to handle food or cigarettes after touching this, since soluble barium is poisonous when ingested. Despite the need for care in their use, barium-containing tubular wires and MMA electrodes have been used for many years without reported damage to welders' health and recent work confirms that the OEL for barium of $0.05\,\mathrm{mg/m^3}$ is probably conservative.

Where particulate fume is removed by filtration or electrostatic precipitation, filters should be washed out using proprietary solutions and never blown out with an air line. Special arrangements may be needed for the disposal of the residues if these contain toxic materials.

## Gases

Gases evolved during welding are often ignored because they are invisible, may be odourless and are difficult to detect without special equipment, but they can represent hazards to health. In welding with metal-cored wire using argon-rich gases, especially at high currents, detectable levels of ozone are generated. Pulsed arc welding can lead to high ozone levels because of the high peak currents used. Ozone is produced by the action of ultra violet radiation on oxygen in the air, so it can be formed at an appreciable distance from the arc and is difficult to control with local extraction systems. Its OEL of 0.1 ppm can be exceeded in confined spaces and even open shops unless care is taken to provide good general ventilation. Processes which produce large quantities of particulate fume tend to generate less ozone because the radiation responsible for it is absorbed in the arc region. Conversely, welding on materials of high reflectivity, such as stainless steels or aluminium alloys, increases the amount of ozone formed. Ozone, which was once used to purify the air in crowded places such as the London Underground system, is now known to cause irritation of the lungs and nasal passages while long-term exposure may cause emphysema.

Other gases produced during welding with tubular wire are usually present at safe levels. Oxides of nitrogen, with an OEL of 5 ppm, and carbon monoxide, with an OEL of 50 ppm, are unlikely to reach these levels in open shops. Submerged arc welding can generate significant amounts of carbon monoxide and, if the flux is damp, silicon tetrafluoride, but these normally disperse harmlessly since the operator is remote from the arc. Carbon monoxide can also be evolved from the breakdown of paints and coatings. All arc welding processes can produce phosgene (OEL 0.1 ppm) if chlorinated hydrocarbons used for degreasing are not properly removed. Phosgene has no smell but is usually preceded by dicloroacetyl chloride, which smells of chlorine and must be taken as a warning of the impending danger.

The responsibility for maintaining safe working conditions lies with the employer, who must set out proper working practices and train personnel in their use. General guidance is available from welding manufacturers but where conditions are more hazardous, for example when welding in confined areas such as inside storage tanks or ships' double bottoms, dangers not only from fume but also from radiation and noise multiply and specialist advice and measurements should be sought.

# References

1 British Patent Application 3762, Feb 14, 1911.
2 'Arc welding in hydrogen and other gases', P Alexander, *General Electric Review*, 1926 *29* (3).
3 'The position of welding in the UK and its relation to the economy', M H Scott, Members' Report R/RB/13/71, The Welding Institute, Cambridge, 1971.
4 'Specification for carbon steel electrodes for flux cored arc welding', American Welding Society Specification AWS A5.20–69, 1969.
5 'Notch toughness of commercial submerged arc weld metal', S S Tuliani, T Boniszewski and N F Eaton, *Welding and Metal Fabrication*, 1969 *37* (8) 327–329.
6 British Patent 858 854, Mar 29, 1957.
7 British Patent 1 510 120, Nov 15, 1974.
8 'Arc characteristics and metal transfer for flux-cored electrode in GMA welding', F Matsuda, M Ushio, T Tsuji and T Mizuta, *Transactions of JWRI*, 1980 *9* (1) 39–46.
9 United States Patent 5 003 155, Mar 26, 1991.
10 'Self-shielded arc welding', T Boniszewski, Abington Publishing, Cambridge, 1992.
11 'Vapor-shielded arc means faster welding', R A Wilson, *Metal Progress*, October 1960.
12 'Mechanised open arc welding with cored electrodes', I K Pokhodnya and A M Suptel, *Avtomatecheskaya Svarka*, 1959 (11) 1–13.
13 United States Patent 3 767 891, Oct 23, 1973.
14 Japanese Patent 62 183998, 1987.
15 British Patent 1 177 993, Nov 21, 1967.
16 United States Patent 3 466 907, Sept 16, 1969.
17 'PP-An3 flux-cored wire for welding low-carbon and low-alloy steels at high currents', I K Pokhodnya and V N Shlepakov, *Avtomatecheskaya Svarka*, 1964 (1) 61–66.

18 'Welded joint design', J G Hicks, Granada Publishing Ltd, St. Albans, 1979.

19 'Welding Handbook', American Welding Society, Miami, Fla, Eighth Edition, 1991.

20 'Switch to metal cored electrodes helps crane fabricator improve productivity', *Welding Journal*, 1992 *71* (6) 75–77.

21 'Control of melting rate and metal transfer in gas-shielded metal-arc welding, Part II', A Lesnevich, *Welding Journal*, 1958 *37* (9) 418s–425s.

22 'Welding characteristics of a new welding process', H Imaizumi and J Church, IIW Doc. XII-1199-90.

23 '$CO_2$ welding with flux cored wires', G R Salter, *British Welding Journal*, 1968 241–249.

24 'Suggested explanation of hot cracking in mild and low alloy steel welds', J C Borland, *British Welding Journal*, 1961 *8* (11) 526–540.

25 'Weldability of steel', K G Richards, British Welding Research Association, Nov 1967.

26 'Deoxidation practice for mild steel weld metal', D J Widgery, *Welding Journal*, 1976 *55* (3) 57s–68s.

27 'Mechanical properties of a mild steel weld metal deposited by the metal-arc process', J M Wheatley and R G Baker, *British Welding Journal*, 1962 *9* (6) 378–387.

28 'Temper embrittlement in steel weld metals containing titanium and boron', A O Kluken and O Grong, Conference Proceedings '*International Trends in Welding Science and Technology*', Gatlinburg, Tennessee, ASM International, 1993, 569–571.

29 'An add-on computer, specially developed hardware and software for documentation and quality control in welding', J Hoejgaard, J Bille and N E Andersen, Proceedings '*Third International Conference on Computer Technology in Welding*', Brighton, The Welding Institute, 1990.

30 'Direct arc sensing adaptive control of MIG welding robots', J R Crookall and M L Philpott, Proceedings '*3rd SERC Robotics Initiative Conference*', University of Surrey, 1984.

31 'Arc hydrogen monitoring for synergic GMAW', T D Manley and T E Doyle, Conference Proceedings '*International Trends in Welding Science and Technology*', Gatlinburg, Tennessee, ASM International, 1993, 1027–1030.

32 '$CO_2$ welding', A A Smith, British Welding Research Association, Cambridge, 1962.

33 'Welding and Cutting', Peter Houldcroft and Robert John, Woodhead-Faulkner, Cambridge, 1988.

34 'Advanced welding processes', J Norrish, Institute of Physics Publishing, 1992.

35 'Present status in the use of cored wires for arc welding worldwide', R Boekholt, IIW Commission XII Project, 1992.

36 'Economics of small diameter, flux cored electrodes', C Zimmerman and D Schmerling, *Welding Journal*, 1983 *62* (4) 41–44.

37 '"Cost savings of 30%" compared with solid wires', *Welding and Metal Fabrication*, 1984 *52* (9).

38 'Rebuilding the blast furnace shell at Redcar', A Gifford and D Baker, *Metal Construction*, 1987 *19* (5) 268–273.

39 'Robot arc welding "doctor" prescribes metal-cored wire', *Metal Construction*, 1986 *18* (9) 534–536.

40 'How Volvo BM exploits metal-cored wire for robot welding – major savings reported', *Welding and Metal Fabrication*, Mar 1987.

41 'Flux-cored electrodes help reduce containership costs 50%', *Welding Journal*, 1978 *57* (6) 36–37.

42 'The use of cored wires in production', G Mathers, Welding Institute Seminar *Cored Wire Welding Consumables*, Telford, 1989.

43 'Continued development brings wider applications for Innershield' W I Miskoe, *Metal Construction*, 1983 *15* (12) 738–740.

44 'Metal-type flux-cored wire for carbon steel', Y Sakai, G Aida, T Suga and T Nakano, IIW Doc XII-1131-89, 1989.

45 'The price of MIG welding', B P McMahon, Conference Proceedings *'Economic Aspects of Welding Technology'*, The Welding Institute, April 1971, 27–37.

46 'Economic efficiency – terms, criteria, requirements', G Aichele, Conference Proceedings *'Economic Aspects of Welding Technology'*, The Welding Institute, April 1971, 227–235.

47 'Difficulties encountered in applying economics to welding', J Docherty, Conference Proceedings *'Economic Aspects of Welding Technology'*, The Welding Institute, April 1971, 214–219.

48 'Self-adjusting welding arcs', J C Needham and W G Hull, *British Welding Journal*, 1954 *1* (2) 71–77.

49 'A new, low-spatter arc welding machine', Elliott K Stava, *Welding Journal*, 1993 *72* (1) 25–28.

50 'Offshore company improves productivity with FCW', J Lukkari and K Mäkalä, *Welding Review*, 1989 *8* (3) 186–188, 196.

51 'Welding in Wärtsilä Marine Industries Inc', J Gustafsson, M Heinäkari and K-E Stenroos, *Hitsausteknikka*, 1989 *39* (4–5) 17–24.

52 'Flux cored electrode developed for zinc primer-painted steel plate', M Kamada, Y Kanbe, T Suzuki and S Maki, *Welding Journal*, 1993 *72* (3) 49–54.

53 'Application experience of flux-cored wire welding, Part II: Cater-

pillar Tractor Co Ltd', R Cable, National Seminar '*Flux-Cored Wire Welding*', The Welding Institute, September 1982.

54 'Design, construction features on Gas del Estados new line', F di Minico, *Pipeline Industry*, November 1987.

55 'Welding district heating pipelines with flux-cored wire', S Due and O Westin, *Welding Review*, 1989 *8* (3) 197–198.

56 'Mechanised welding methods in pipeline construction', R Boekholt, Proceedings *Pipeline Technology Conference*, Gent, Royal Flemish Academy of Sciences, 1990, 2.29–2.73.

57 'An assessment of flux cored wire welding type 316L stainless steel', T G Davey, T G Gooch and J L Robinson, *Metal Construction*, 1987 *19* (8) 431–435 and *19* (9) 545–553.

58 'Hazards from welding fumes', Leaflet 236, The Welding Manufacturers' Association, 1985.

59 'Safety in welding and cutting', Publication Z49-1, American Welding Society, Miami, Florida.

60 'Assessment of exposure to fume from welding and allied processes', *HSE Guidance Note EH54*, HMSO, 1990.

61 'Occupational exposure limits', *HSE Guidance Note EH40*, HMSO, revised annually.

62 'Guidance Note 15 from the Health and Safety Executive', HMSO, 1978.

63 'Cancer risk in welding', W Zschiesche, IIW Doc. VII-1654-92.

# Index

acicular ferrite, 69–71
alloyed wires, 29
aluminium in tubular wires, 26, 29
American Welding Society, 10, 13
   Specification A5–20, 13, 26
arc instability, 72–74
arc stabilisers, 12–13
argon mixtures, 20, 22, 44–45, 57–58
austenitic steels, 29–31

back-gouging, 41, 71
barium compounds, 27
basic wires, 13–16
basicity, 14
British Standards
   BS 449, 8–11
   BS 5135, 60, 61, 65
   BS 6693, 61
   BS 7084, 61, 63
burn-off rates, 45–47

carbonates, 13, 26–27
CEN (European Committee for
   Standardisation), 10
$CO_2$, 20, 22, 30, 44, 53, 57–58
cold cracking, 60–65
cold laps, 6
contact tip, 8
   excessive wear, 74
copper, 60
creep-resisting steels, 24, 29
current density, 4

deoxidation, 18, 25–27, 34, 35, 55–56
deposition rates, 4, 21, 45–47
dilution, 71
dip transfer, 7, 8, 10, 15, 53
drawing as a production method, 33–34

drawing soaps, 7
duty cycles, 3

electrode extension, 11
embrittlement, 68–72
European Standards
   EN 1011, 60, 65
   EN 439, 66
   EN 729, 75
   EN 758, 23, 63

filling ratio, 23, 31, 35–36, 65
fluorides, 13, 25–26, 28

guns, 8

heat input, 69
high strength steels, 29
hydrogen, 7, 14, 15, 17, 29, 58–65

joint preparation, 41–43, 53

K-preparation, 68–69
Kjellberg, Oscar, 1

lack-of-fusion defects, 7, 52–53
lithium compounds, 28
lubrication, 37–38, 73–74

microstructure, 68–71
MMA welding, 1, 2, 3, 6, 21, 42, 61

nitrogen, 14, 55, 57, 66–67
nozzle diameter, 57

oxygen, 13, 14, 16, 55

phosphorus, 59–60, 66

polarity, 15, 23, 25
porosity, 7, 17, 55–59
positional welding, 15, 16–17, 19, 27, 41, 53
power sources, 8, 15, 48–50
primers, welding over, 15, 21, 23, 43
production methods, 31–35
productivity, 3–5, 41–51
pulsed arc welding, 48–50

robotic welding, 44, 50–51
rolling as a production method, 32–33, 35
rutile, 16, 27–28
    wires, 16–18, 29, 41–47, 77–78

seamless wire, 34–35, 57
segregation of alloying elements, 65, 66
self-shielded wires, 25–29
semi-automatic welding, 2, 8, 76
sheath materials, 31
shielding gases, 11, 20, 44–45, 57–58, 61–, 62, 66–67
shipyards, 21, 49
shroud, 8, 53, 77
slag, 12, 24
solid wires, 21, 53
solidification cracking, 14, 59–60
spray transfer, 8, 16

stainless steels, 29–31
standoff, 11
stickout, 11, 17, 26, 45, 53, 61–62, 75, 77
strain ageing, 71
stress relief, 43
submerged arc welding, 11, 30
sulphur, 14, 15, 59–60, 66
surfacing wires, 30
synthetic titanates, 17

temporary backing systems, 41, 42
terminology, 8–11
titanium dioxide, 16
torches, 8
toughness, 14, 17, 47
training of welders, 77–84
travel speeds, 21

underbead profile, 21, 22
undercut, 77

voltage, setting up, 77–78

welding monitors, 75
welding speed, 47–50
wire feed speed, 5, 53
wire feeding difficulties, 73–74
wire sections, 35–37

Printed and bound by CPI Group (UK) Ltd, Croydon, CR0 4YY

08/05/2025

01864838-0007